LIFE AND TALES
OF A LOCKSMITH

Our Family Cottage from 1965-1988

LIFE AND TALES OF A LOCKSMITH

K. W. "Bob" Sidbotham

HiP
HISTORY INTO PRINT

First published by
History into Print, 56 Alcester Road,
Studley, Warwickshire B80 7LG in 2005
www.history-into-print.com

ISBN 1 85858 306 3

A Cataloguing in Publication Record
for this title is available from the British Library.

Typeset in Caslon
Printed in Great Britain by
The Cromwell Press

CONTENTS

ACKNOWLEDGEMENTS

Over the years, I have incurred many debts to family, friends and acquaintances in compiling my autobiography. To name them all would almost need another book, but for their special contribution my thanks go to Professor Owen Ashton and Dr. Michael J Mulligan.

The greatest debt, naturally, has been to members of my family: my daughter Carol for all the typing, my son-in-law Pete for his computer skills and reprinting of photographs and my brother Geoff for all the art work. But, above all, it is to Ann, my soul-mate and partner through thick and thin. This book is therefore respectfully dedicated to her in loving memory.

Bob Sidbotham, Gnosall, Stafford

May 2004

INTRODUCTION

Firstly I ask myself "can I write a story?" about my life as a locksmith, "will anyone want to read it?" I will never know unless I try, and if I don't try I shall not know anyway, neither will I have the pleasure and enjoyment of putting my experiences into writing, so that younger generations will have the opportunity of reading, if they wish to do so.

It seems a great pity that many ordinary people like myself, depart from this world without leaving a written trace or pattern of how they conducted their lives in business or socially.

However the first part of my story is about my life as a locksmith working in a small family business, which belonged to my father. The second part begins during May of 1965, when I left the industrial Midlands and went to live in a derelict cottage in a remote area of Shropshire where I continued to work my trade.

My story is a true story, and has a humourous side to it. It tells of the pitfalls, and aggravations, which I have tried to make interesting without creating boredom to the reader. Also I have avoided the use of locksmith's jargon. It is not a technical writing, it is a story easily understood.

Locksmiths are like most craftsmen, a breed of people dedicated to their skills, to which patience is a virtue. The making of hand made locks for safes, and strongrooms, has all but died out. I think never to return. Gone are the days, thank goodness, when men stood at the vice for twelve hours a day, six days a week, and were inadequately paid for their labours. I have now retired. My whole working life, and a great deal of my private life has been dedicated to the lock trade, and the small business I operated. I have written this story for my granddaughter, Victoria, (Bunji) whom I am sure will one day read it.

PART I

Early Days

Chapter 1

THE LIFE AND TALES OF A LOCKSMITH

It all started back in 1948. I had been demobbed from the army. Like many thousands more ex-servicemen, I had come home to make a new life, and help to re-build our country after devastation during the war years. Before I enlisted, I was serving as an articled student to a firm of auctioneers and estate agents in Birmingham. After my release from the forces I could not afford to continue in this field. However to come to the point, my father had started his own business during 1946, he was a locksmith, the old type, a craftsman, like his father before him. At this time he had one employee, and it was on the cards that I was to be number 2. In fact whilst I was away in the forces he would regularly say to mother "he needn't cum 'ome if he ain't gonna wirk for me" so after a lot of thought, a lot of talk I decided I would give it a go, although we had yet to discuss money, well that was going to be an obstacle, I knew his feelings about money, never spend a penny if a halfpenny will do. "So", said the old chap, "I'll give yer three pounds one shilling a week, that's the union rate and the same for holidays, one week with pay, tek it or leave it." "OK", I agreed, I knew he was one of the best locksmiths around; I was going to learn the hard way, but the right way. I had seen the place where I was going to work, enough to put anyone off, a dark, dreary ancient workshop in Wolverhampton. Father rented it for eight bob a week, and sub let half of it for four bob. The downstairs workshop, I called it the "Black hole"; there was a pile of rusty lock pressings, which stray cats got through small broken windows and piddled upon, and mucked on a heap of coal slack in the far corner.

There was a set of old loft type stairs leading to a trapdoor, and into a small workshop, it housed a bench, three old leg vices, and other bits of antiquated machinery. Another room was through an old door made from bits of odd timber with cardboard nailed to it. You may think it was a fair sized building, but it was hardly big enough to swing around the stray cats it harboured. After I had been there a few days I noticed a lot of mouse droppings, so I decided to go back one night and investigate. There were two cups on the bench, low and behold, mice were perched firmly on the brim of both, and sucked away happily at the leftovers or dregs, the cats were most welcome from then on, and were given a daily milk ration. I remember another night when I was going to close the upstairs windows, there was a scuffle, one of the cats jumped through the open window, fell to the ground, and away across the patch. During my first couple of weeks I was put to work on an old drilling machine, which looked as if it had come out of the Ark; it was driven by overhead shafting with flat balata belts, there was a makeshift attachment; it was a wartime dried milk tin, which the old chap had made into a guard and tied it to the driller with a piece of string. However, after I had been drilling for a couple of days the string broke, down came the milk tin guard, my hair was caught in the shaft, I was all but scalped. There were no sympathies in those days, "that'll tek some o' the mad blood out a y'a" shouted the old fella, and it did, it also took twelve months for my hair to grow again.

I was a slow learner, but very determined. To me the place where I was to earn my living was nothing more than a shambles, not the easiest of places for an ex-office lad to adjust himself to. However I was there to learn lock making not to criticise and I didn't want another experience like the one I had just had with the old driller.

I think Father felt sorry for me at that time. "Come on Bob", he said, "it's time we got down to some serious business", I'll show you how a lock is made". "Go and get the castings for the six lever block lock and we'll go through it together". "Make sure there's no blow holes in them".

The castings were made of brass from our own patterns, and the locks were fitted on safes and strong room doors made for the Birmingham Municipal Bank. The maker of the equipment was Samuel Withers & Co Ltd, of Barton Street, West Bromwich.

"We are going to make one of these", said Father.

I was to clean and fettle the castings, then Father would level and grind them. He used to say levelling was very important and that it was the foundation on which the hand made lock was built.

I drilled the fixing holes (picture 1) in the lock body. Then I drilled, tapped and countersunk the holes for the fixing of the top plate. Drilling the keyhole was very important this was where Father's teaching paid off. If the castings weren't level, and flat before drilling then the keyholes would not be vertically in line (picture 1). As the old lockies used to say "yer kays are drunk".

The keyhole slot was put into the top plate only, using the hammer, a leg punch and die, which was part of the locksmith's tools (sketch 1). Later on Father bought a Fly Press and we made a press tool for that job.

Father fitted the bolt in the bolthole after which he rivetted the footed drillpins firmly into their countersunk holes. "Like all jobs", he said "there's a right and a wrong way" (picture 2). For instance, if we were to countersink too deep for the foot of the drillpins they would not tighten. Also if they were hammered too heavily the foot could break off.

There were two drillpins in the lock, one to run the bolt, and also to act as a pivot pin for the six levers. The other was a pressure point for the tension of the six springs. The drillpins were gripped in a pair of stocks held in a leg vice and rivetted into the lock after which they were cleaned off with a cutter. The stocks and cutters were usually made by the locksmith and were part of his own personal tool kit (sketch 2).

THE BOLT (picture 3)

Father had fitted the bolt and filed the range slot which enabled it to run freely on the lever pivot pin. At this point the thickness of the bolt tail had to be precise and was filed and gauged with the use of a Vernier calliper before the flat stump was rivetted into the bolt tail. Working clearance between the top lever and the underside of the top plate was important. There were four fixing holes in the lock and it was secured to the safe with four 5/16" Whitworth bolts, hence the importance of having the correct tolerance under the top plate. The shoot of the lock bolt was determined by the length of the range slot in the bolt tail.

Our next job was to drill a small hole in the bolt tail and rivet in the flat stump. To obtain the location for the hole for the stump we put an ungated lever (See Picture 4) on the pivot pin keeping the bolt in the unlocked position. We marked it in the unlocked chamber with a centre punch, drilled the hole and fitted the flat stump. After a few very important adjustments to the range slot and the flat stump which was known as the pitching we were ready for the levers and keys.

Traditionally key blanks were made from malleable iron castings (sketch 3). The key makers, or filers as they were sometimes called, bought the Annealed key castings from the iron foundry. They made them into respectable key blanks almost ready for the locksmith's use, although in most cases the locksmith would have to make adjustments with the file to obtain a steady fit in the keyhole of the lock.

Like locksmiths most keymaker's machinery was antiquated homemade, and out of date. Father picked out a couple of six lever key blanks. He put the levers on the lock and he went to the machine to cut the keys.

Our key machine was made up with bits and pieces from a couple of old lathes, it had a sliding saddle which was essential for sawing keys It also had a notched gauge on the saddle which was used for differing the keys.

The lathe was worked with a single saw. Every step on the keys had to be gauged and rounded with the file on what was called a key horse which

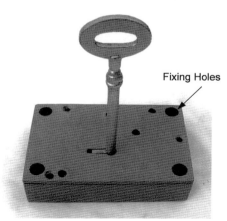

Fixing Holes

Picture 1 shows the Key, vertically in line
along with fixing and location holes

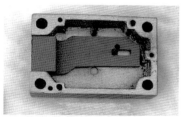

Picture 2 shows the Bolt fitted into
the lock and two Drill Pins

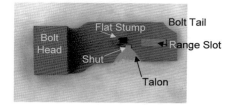

Flat Stump   Bolt Tail

Bolt Head   Range Slot

Shut

Talon

Picture 3 shows the Bolt in detail

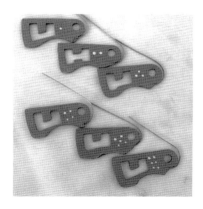

Picture 5 shows six gated Levers

 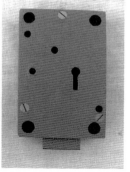

Locked Chamber   Unlocked Chamber

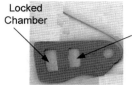 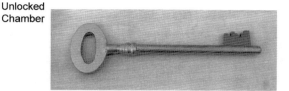

Picture 4 shows an
un-gated Lever

All of the above pictures show a six Lever, Down Shoot Safe Lock.

A working model made by the Author from laminated cardboard other than Lever Springs
(Phosphor Bronze), Screws (Steel) and the Key Shank (Wooden Dowelling)

had a needle gauge attached to it. These tools were also made by the locksmith and were part of his tool kit (sketch 4). To quote the old lockies: Get "yer kayoss in the vice and round up yer kays".

Laterally, the cutting of the keys was quite accurate because father always had a sample lock and could make random checks as to whether the machine had moved or not (sketch 5).

We worked with twelve sets of lock parts on a tray, also twelve pair of cut keys on another tray. The amount of times that the locksmith would pick up the twelve pair of keys when gauging the steps could be well in excess of 168 minimum, depending on how accurate the keys were cut. The operation would take a skilled locksmith approximately forty to fifty minutes to complete.

During 1956 we moved on by cutting keys the new way which was known to the trade as 'Block or Gang Sawing', hence the gauging of key steps for production lines became obsolete. Accuracy was the name of the game.

At this time the manufacture of key blanks made from fabricated mild steel was being introduced. Although not without scepticism, often the key bit or bow would break off, especially whilst cutting the keys under machine pressure.

The safe makers could not afford to have key bits breaking off behind locked safes or strong room doors, so they expressed their doubts and did not adapt themselves to the new welded keys until they were improved.

To shape the bolt talon and fit the levers is the next step. The lock is set up in a jig with both, the bolt and key in their working position, the top plate is removed so that the key can be used in the jig for fitting the levers (sketch 6). The bolt talon is important and it is related to the bolt step on the key and is shaped into the bolt tail to suit the key. Making sure that the lock bolt is in the fully locked position we mark it to the height of the bolt step. Then we draw the bolt back fully into the unlocked position and repeat the process. After which, we remove the bolt from the lock body in the jig. Put it in the vice and file it with a half round file, evenly from the centre of the two marks, checking all the time with the key until we have formed the shape of an arc in the bolt tail.

Finally a small shut is filed in the arc and we have a perfect talon. The small shut was also known in the lock trade as the throating.

THE LEVERS (pictures 4 & 5)
There are two chambers and a pivot location hole in the levers. One chamber for the locked position. The other is unlocked. The centre bar between the two chambers is known as the gating.

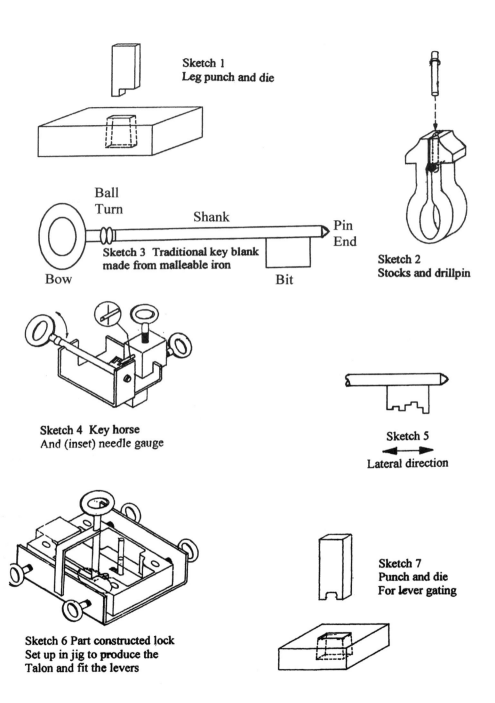

Sketch 1
Leg punch and die

Sketch 2
Stocks and drillpin

Ball
Turn
Shank
Pin
End
Bow
Bit

Sketch 3 Traditional key blank
made from malleable iron

Sketch 4 Key horse
And (inset) needle gauge

Sketch 5

Lateral direction

Sketch 6 Part constructed lock
Set up in jig to produce the
Talon and fit the levers

Sketch 7
Punch and die
For lever gating

**Key steps showing workable and accepted differs for six
lever safe door locks costing 6s.1.1/2d.(30.5p) each in
1948 with two keys and were guaranteed to differ.
There were five lifts from high lift to dead lift
A single saw was used for key cutting which became
obsolete during the late forties or early fifties**

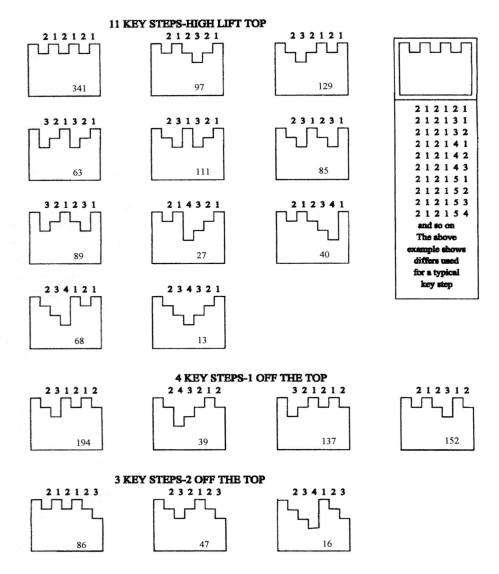

11 KEY STEPS-HIGH LIFT TOP

2 1 2 1 2 1 — 341	2 1 2 3 2 1 — 97	2 3 2 1 2 1 — 129
3 2 1 3 2 1 — 63	2 3 1 3 2 1 — 111	2 3 1 2 3 1 — 85
3 2 1 2 3 1 — 89	2 1 4 3 2 1 — 27	2 1 2 3 4 1 — 40
2 3 4 1 2 1 — 68	2 3 4 3 2 1 — 13	

2 1 2 1 2 1
2 1 2 1 3 1
2 1 2 1 3 2
2 1 2 1 4 1
2 1 2 1 4 2
2 1 2 1 4 3
2 1 2 1 5 1
2 1 2 1 5 2
2 1 2 1 5 3
2 1 2 1 5 4
and so on
The above
example shows
differs used
for a typical
key step

4 KEY STEPS-1 OFF THE TOP

2 3 1 2 1 2 — 194	2 4 3 2 1 2 — 39	3 2 1 2 1 2 — 137	2 1 2 3 1 2 — 152

3 KEY STEPS-2 OFF THE TOP

2 1 2 1 2 3 — 86	2 3 2 1 2 3 — 47	2 3 4 1 2 3 — 16

With the lock in the locked position and the first lever assembled on the lock we proceed as follows: keeping the lever in contact with the first step of the key, we turn the key to its highest point until the flat stump comes into contact with the bar in the lever, its position is then marked off and slotted with a punch to produce (The gate). The accurate part is completed by filing the belly of the lever until the flat stump passes through the gate, perfectly. Then the spring is fitted and we are ready for the next lever (sketch 7).

A similar process is repeated in the gating of the next five levers using the key at all times. After which we remove the lock and key from the jig, put on the top plate and screw it up. The lock is finished off. The keys are polished and it's ready for use.

The end of my first lesson 1948 (see accompanying diagrams and photographs).

There were no dust extractors, brass grindings filled the air, and glittered like gold in the rays of the sun, it was all over the place. I remember very well, when the other bloke that worked for us, shouted "look at the brass dust shining in the sun gaffer", "ya'll atter get a fan", "Well" said the old fella, "the sun'll be gone in soon then ya wo see it". But eventually we did have a dust extractor, it was partly home made but it worked. Needless to say, like the driller, part of a dried milk tin was used. "Them tins is useful", the old chap would say, "kept em I av, ad em a long time, never throw em away". The old fella was a very clever man, and an artist when it came to make do and mend; also he was a great improviser. I suppose it's very easy to criticise someone who is trying to make money and build up a business, especially if one hasn't any money to start with.

Anyway to get back on track again, it was winter, January 1948, when I was learning how to use the file, this was one of the most important tools used in our trade, and was vital for the making of locks and keys by hand. In those days we would arrive at work at about seven thirty in the morning. The old chap would go on his push bike, and me on foot, in the winter we would light the fire, usually on an uncleaned out stove with a pile of long standing ashes in front of it, if I tried to warm my hands before handling cold metal, the old man would say, "what are ya standin there fer, pick up a file and file some castings, that'll warm ya up, and ya'l learn how to file at the same time". Anyway after a few months, I was getting used to being there and doing a bit of useful work, but I had drifted into working extra time without being paid for it, so I approached the Dad about some extra cash, well I had to ask or I wouldn't have got anything, although I knew it would cause a caffufle, which it did.

So it's overtime pay ya wants is it he said. "Well Dad, I'm working seventy hours a week for just over three quid" I replied. "But that's the way ya learns ma lad", says the old fella. So on that occasion I got nothing; I began to think I was a born loser, because I was only working for peanuts anyway. Although I did have a rise a few weeks later. He was put under fire and who had the finger on the trigger, Mother, that's who, and she didn't ever lose, no sir, not Mother.

So after a few months I was able to save a bob or two, and I bought myself an old motor bike, it was a pre-war machine made in about 1934, a B.S.A. 150cc overhead valve, four stroke engine and in good condition. This was handy for me to get to and from work, and didn't cost much to run.

It was always late at night when I got home from my work, except for Wednesdays, that was my only night off. Ann my girlfriend used to pay for us to go to the pictures in those days because I hadn't ever got any money.

Ann and I had known each other since Junior School days. She lived in the next lane to me, which was only about a quarter of a mile away, so it was easy to meet her on the other nights after I finished work, which was usually about half past ten at night or later. I would leave the workshop at the same time as my father, on the way home there was a long steep hill, and I knew that the old chap would have to push his bike to the top of it, so I would wait for him at the bottom, take a short run with the motorbike in second gear, place my left hand on his back and push him to the top of the hill. I bet his old bike never went so fast!

We worked on Saturdays and Sundays, all hours, when I wasn't working with him I always had a conscience, and I would go back to help him get the work out. Although he never seemed to appreciate this, I think he did. After a while he got a bit better off, and bought himself a second hand pre-war Austin Cambridge saloon car. It was made in 1939 and was in immaculate condition. He was so obsessed with it he sat in it every day on the patch to eat his grub.

Father had taken on a young junior to teach him the trade, he had the making of a good locksmith, but unfortunately he was forced to leave the job when he contracted Tuberculosis, which was a scourge of our time. Whether this was due to our working conditions, we shall never know, because brass dust, and swarf, tumbled through the gaps in the floorboards; it was most hazardous working on the ground floor.

Underwear, shirts, even pillowcases, in fact any fabric turned green. The mixture of sweat and brass was here to stay, no matter how often one washed. The only way to overcome this dilemma was to have modernised the old workshop, and to introduce efficient dust extractors. All these problems were getting me real up tight. I was frustrated, angry if you like, ready to explode, or leave. Anyway out of the blue, before I did blow my top, Father let something slip about buying a small factory in Wednesfield. I could never understand how he stuck the old place so long.

Anyhow things were on the move, he was negotiating to buy. In the meantime he took on another workman and we had reached a large workforce of five including myself and him.

Our new man was a strong trade unionist, a troubleshooter, not because he was a trade unionist, but because he was a troubleshooter as you will read later on. Father didn't like him neither did I, although he was a good locksmith, just what we wanted. I nicknamed him the "Union Reb" or "Rebel".

Chapter 2

MOVING FROM THE BLACK HOLE

It was during the summer of 1951. We were about to move out of what I called "The Black Hole". I say we, as a manner of speaking, I was nothing in the business at the time, other than an employee. Anyway the deal was completed, I hoped that a change of scenery and environment may lead to better prospects for my future, we shall see. My brother didn't work for the Dad, that's perhaps why they got on so well together. So whilst the old chap had taken mother on a touring holiday in Cornwall, brother Geoff was going to move everything from the old place to the new premises.

The Austin Cambridge was highly polished, and they were on their way. So the move took place, *lock*, stock and rubbish, the whole shebang. "Operation Black Hole", had been successful, mission complete. Now although the new factory was small, it was large compared with the old dump. It was a rectangular piece of ground with a boundary wall, there were brick toilets and a wash basin; we never had it so good.

The workshop was of timber construction, which had now become full of machinery and junk. On the Monday after the holidays we turned in for work. Brother Geoff had done a great job, must have sweat his guts out. "Where's 'ee put it all", "ee's mixed it all up", shouts the old chap. "Better sort it out as quick as we can", and start to earn some dough, holidays is over now". So after two or three days scrattin and tattin, we were able to earn a bob or two and brother Geoff was praised for his good work.

After being there a few weeks, we submitted plans to the local council for an extension. It was to be a Nissen type building, ex War Department. The plans were passed, it was delivered, and three men came to erect it. I remember making a comment to one of them, "I've slept in these huts for nearly four years, when I was in the army". The old chap heard and shouted, "well try bloody well working in one fer a change"! So we settled down to work, all went well for some time, at least for a couple of weeks, until we had trouble with the driller, the one from the Black Hole, the one with the milk tin guard, which part scalped me. The Dad had modernised it to save buying a new one, but it was no good, kept stopping every few minutes. So the trade union rebel I talked about was feeling his feet, "Ya gorra get a new driller in 'ere gaffer", he shouts, "can't earn me muney on this ode thing, if ya aye got one fer next wick, om off ta get me someweer else ta wirk".

Anyway the driller was soon replaced, and an electrician came to do the wiring up, and we were quiet once again for a little while. I was pleased to see the old driller go, and on this occasion, had to agree with the Union Reb. It was now early 1952, I was engaged to Ann, and getting married in March of that year, I could claim a tax rebate if I married before the end of the income tax year, which was in the April. Somewhere around this time, the Gaffer, I call him the Gaffer, because he was anyway, looked at his Austin Cambridge and decided it needed a re-spray or a complete face-lift. Now the Dad always had a habit of shouting about what he was going to do. Couldn't keep his own council as they say, neither could the Reb keep his mouth shut, "ya doe wanna be spending muney on them fancy motors", says he, "why doe ya buy another machine"? That set all four of us going, I took the old chap's side this time, so everybody's jabbering, nobody listening, nobody's working, we are all losing money, because we're on piece work.

So the Cambridge went in dock. We purchased an old vehicle to carry us on, it was a van, 1933 Morris Eight, ex Post Office, it was OK, although a bit of a boneshaker, but what could one expect for sixty quid. About fourteen days after we bought the van, back came the Cambridge, like new, by gum it did shine.

Anyway as I have said I was getting married on the 29th March. At that time I could earn about nine pounds a week on piecework, and was able to put in extra hours to enable me to save for our honeymoon. We were going to a hunting lodge in the Forest of Dean at Cinderford. It had been said that we may borrow the car, but it looked so good, would he lend his prized possession, no not the old chap. Anyway I prompted him and he quickly replied, "yo ain't avin my car, ar've just ad it dun up, so tek the ode van". So off we went in the ode van. The next few lines take me a bit off track, but I thought rather funny, when we arrived at our destination on the Saturday evening, our cases weren't with us, "must have left them in the Cambridge", I said. Some of the contents in the case were important, couldn't afford to slip up in those days.

So the first night was a frustrated fiasco, but after digging ourselves out of a foot of snow the next morning we came home for the cases, went back to Cinderford, and had a whale of a time. After a week I returned to the unpleasant grind of Locks and Keys, all was going quite well. We lived with my in-laws for the next eighteen months, during this time the Cambridge was sold, and the Gaffer bought a new Ford Consul. He was real chuffed, as proud as a peacock. I remember they had seat covers to keep the seats clean, cloth on top of the covers, paper on top of the cloth. I remarked, "when are you going to stop Dad? You won't be able to get in the bloody car if you keep going on like this". "Ah but ar've gotta look after it, it's cost me a lot of munny". I was beginning to wonder whether I ought to approach him about a possible interest in the business. However, it didn't seem right at the time, he had just bought a new car, and I had nothing to offer, at least not cash wise, but my knowledge of the trade was improving, and I new that we had to get down to better productive methods if we were going to prosper. However the old chap was a stick in the mud, and didn't want to know about modern methods or spending money, anyway other things were on my mind, moving from my in-laws was a priority, because we wanted to be on our own. My wife Ann had known two old ladies for some years. Both spinsters and costumers by trade, they were in their eighties, and lived on their own in an old Victorian town house in Wolverhampton where we were given the chance of renting the top floor as a bed sitter, and another small bedroom which we could use as a kitchen. Our rent was to be ten shillings a week. "Great, when can we move in Miss Smith", I asked, "Anytime you like" she said, "the weekend", "Yes all right, thank you". So we moved in, built a sink unit, with a bucket underneath, carried fresh water upstairs, sink water down, coal up, ashes down, there were three flights of stairs; toilets were outside at the back. The old ladies were a bit fragile and were pleased to have a man about the house, so we stayed there and were very happy for two years.

Living at the flat was very lucky for us, and plays a very important part in my story, which is to follow.

In those days I used to have the occasional bet on the horses and I remember one Saturday, it was a hot summer's afternoon, we were watching the racing on the television, when there was a tapping on the window pane, it was quite a shock especially when you were three flights up. We looked, couldn't believe our eyes, when a fireman peered in and shouted, "get out, next doors on fire". So we scarpered double quick. A hot day in more ways than one. A few weeks later I was walking home, it was in the late evening, when in the reflection of a street lamp, on the wet pavement, I saw what I thought was a bank note, I picked it up, and found three soggy one pound notes crumples together. It was a good find in those days. I asked myself, should I save them, no, easy come, easy go.

Our 1933 Morris Minor 8 Horse Power van, ex-Post Office. Used for lock deliveries. Cable front brakes working around brass pulley wheels. Hydraulic brakes were introduced by Morris the following year. Purchased second hand in 1951 for £60!

So I had a bet on a horse, three pounds straight win. I always remember the horse's name, it was 'Coronation Year', and it won at a good price fourteen to one. A few days after this, I saw Harry. Now Harry was Miss Richard's brother, she was the other old lady who lived in the house, Harry used to come every week to see her. He was a betting man, in the know, if you know what I mean, so I told him about my bit of luck. "Oh good, I'm very pleased fer ya Bob, I've got a good un cumin fer next week", he said, "straight from the orse's mouth, as they say". "Can't give it yer now, gotta keep the odds high, but I'll see you before the race". "OK, thanks Harry", cheerio. The next week on the night before the great day, Harry came, "Bob" he calls, "can ya spare a minute". "Yes Harry, I'll be right down". He thrust a bit of paper in my hand and on it was written the word 'Preloan'. "It's too good to miss and don't tell everybody" said Harry. "The owner is a friend of mine, and tomorrow night his chauffeurs going to pick me up in the Daimler", you can come as well if yer like". "OK Harry, thanks a million". "Don't forget, Bob put yer shirt on it", he shouted, and away he went. I was a bit doubtful, but Harry wasn't the type of bloke to give anybody a load of bull, especially me, so I took him seriously. I mustered all my cash, everything I had in the world, and rooted in pockets, tins, anywhere to find money for the gamble I was about to take. However, there was one exception, the old man was going to be the one fly in the ointment. He knew I was going to bet, and he owed me some back pay. I wanted it; he wasn't going to let me have it, no way. I pestered him but he wouldn't budge, not one inch. "I ain't gonna give it ya to bung on them 'orses," so I didn't get it. But the horse won, and I was in the money, over five hundred quid! The others wished they had backed it, even the old man said he wished he had bet. "I never thought it would be that good", he said "I am a bloody fool", that was the only time I heard the old chap call himself a fool. I was over the moon.

I went home that night, Harry and the chauffeur came in the Daimler to pick me up. I borrowed a bob or two, and off we went for a booze up. The next day the bookie paid me cash on the nose.

For the next three weeks following my win, I gambled on the fixed odds football coupon, and won one hundred pounds each week for two weeks. The third week I cut the stake and won fifty pounds. After this, the Dad stuck his oar in, "why doe ya pack it p before ya give it all back ta the bookie." I valued his advice, it was good, and turned out for the best. "OK Dad, but I am having one last bet." Fifty quid straight win on a horse called 'Nicholas Nickleby'. "It's a donkey," he said "it'll never win a race in its life." "All right I'm still going to bet, and if it ducks that's the finish." It ducked, I packed up all form of gambling whilst I was well in the money, except for two five pound bets which I had over a period of the next thirty seven years. Although I was tempted one day later in life, when having been called out to open a safe for a bookmaker in Birmingham, my resolve almost slipped as you will read later on.

I was getting on quite well in the lock trade. I had all my winnings, but what was I going to do with my money, I needed to invest it, buy a house, or even try to talk to the old chap about going into the business.

Ann and I had agreed, the business must come first, so I approached the old fella, it wasn't easy. "I've got all my winnings," I said, "If I put them into the firm can I become a partner, what do you think Dad?" "Well yes, ah, OK, but, I'll ave ta tell your mother when I get ome, and see what she's got to say."

The next day we discussed the subject in more detail, but it was not to my advantage. I was invited to put my savings into the business with a view of helping to improve our methods of production. Also it would enable us to employ female labour which would increase output, and introduce an assembly line on some of our locks, especially the ones where there was a demand for larger quantities. All this was to be done at my own risk and expense, with no agreement. Father wasn't going to lose his money, however, if I was to be successful I would get a partnership. I couldn't get any improvement on this, so I took it on.

The first line of production I was going to develop was for a lock, which was already being fitted to safes and strong room doors. We were making it by hand, the old fashioned way, and our selling price was six shillings and three half pence each, three pounds thirteen shillings and sixpence per dozen. A piece work price for making these locks was eighteen shillings a dozen, the maximum quantity I had ever known any one workman make in a week, working a total of fifty five hours was one gross (144). His wages were ten pounds sixteen shillings, so it was obvious that money was essential for progress to enable us to survive. The trade was also highly competitive especially when dealing with safe makers, who tried to almost beg the locks for their safes.

I bought machinery, made press tools, and worked all hours as many as ninety in some weeks. Home life had virtually gone, other than supper and bedtime. I was progressing OK, there were many pitfalls, and much opposition from the old chap, the old and the young did not mix in my view, especially when related. However my money was running out and this caused problems between the Dad and myself. Almost all our eggs were in one basket with a firm of safe makers Samuel Withers & Co. Ltd at West Bromwich. They were tough people to deal with and to raise the prices of out locks was almost impossible.

It was now 1954, our workforce was up to a total of six including ourselves, Union Reb had gone, got himself a better job with more money, he said. Financially we were getting worse off. Our difficulties were increasing, to pay the wages was becoming a problem and to make the locks the old fashioned way by hand was almost impossible because our customer was demanding larger quantities, and getting less. The reason for this was that we could not make the tools quick enough for the assembly line which I was trying to create, neither could we afford to put them out to be made as I had previously done.

So as we sold our locks, we had to be paid for them immediately. This meant giving large discounts for the privilege, our profit margins were very low, and we were in a back-to-the-wall position.

I had made good progress in my work, but I couldn't move any further forward. I was at a standstill and once again under fire from the old chap. I remember him saying, "you've got us into a fine bloody mess with them big headed ideas a yours". I remarked "how can I go on any further without money, I've used it all?" He hadn't got any either, because he had just bought himself another car, I approached him, "why can't you get a bank loan to help to get out of this worry?" "No bloody fear," he said, "If I got one a them, you'd want another next wick and I'd atta sell me ouse to pay fer um." As I look back I can understand how he must have felt.

However, there was one more machine I had to have which would enable me to accomplish what I had set out to do. It was a thirty-ton power press. Without a power press I could not put into practice the tools I had made, and bought from toolmakers. So it was vital that I got one somehow.

The author hard at work.

Temporarily, the position went from bad to worse, our telephone rang non-stop for a couple of days, neither the old chap or myself would answer it, because we were sure that it would be our customer wanting the locks, which we had promised many times and failed to deliver. The next thing was the arrival of a telegram, it read as follows; "Why don't you answer your telephone, ring me straight away," signed D. Withers, Samuel Withers and Co. Ltd., West Bromwich. Father rang him, and Dennis Withers went bonkers, blew his top completely. I've got safes and strong room doors waiting for tuppenney 'apenny locks to be fitted, and eighty men sitting around on their arse's, what are you going to do about it?" "I'll be over this week with as many locks as we can get done," said the old fella. "Promises" says Dennis, "Ya promises are like pie crusts, and what about that production line which your Bob's doing?" The Dad lapsed into his Black Country lingo, as he usually did when he got uptight about anything. "It ain't nothing ta do wi me, ee's got these new fangled ideas." "Well put him on the phone, let me have a talk to him." This was an opportunity that I couldn't afford to miss, so I told him about my problem regarding the power press. "I think both of you had better come and see me, and bring all the locks you have ready when you come." So at the end of the week, on the Friday afternoon we trundled off to West Bromwich. Now Dennis Withers was a tough businessman, but if caught on the right foot could be reasonable. I told him that I had run out of funds and couldn't go any further with the job. "Well Bill," that was fathers name, "why don't you take up a loan, Bob's got to finish the work he's set out to do or he'll lose everything." "No fear, I ain't avin any of them bank loans, 'ee might lose it, the way 'ee's going on," said the old fella. "Well," said Dennis once again, "give him a chance to finish the job." "No I'm not tekin any bank loans and that's that." Dennis looked puzzled; he couldn't understand why the old chap wouldn't take on a loan after I had gone so far forward with my work. "How much is it going to cost to buy this press?" he said, I was quick to answer, "I've seen just what we want at four hundred and fifty pounds." "All right then, I will lend you five hundred pounds, but there will be one condition, my solicitor will form a Limited Company for you, and we will be debenture holders until you repay the loan, and don't forget I want all the locks you can produce as soon as possible until after the loan's repaid." I waited for father to comment. "All right" he said. Why did he have a change of heart? Perhaps he felt a bit guilty, or could it be that the loan from Dennis Withers was interest free, I never knew and didn't ask him. It was just what I wanted, to be in the business and a Director of the new company. So our company was formed. Father and I were working together, at least for the moment. He had previously been trading under his initials, W.B.S. Safe Locks; we were now W.B.S. Safe Locks Ltd.

Samuel Withers & Co Ltd. were debenture holders as arranged, and our loan was repaid by deductions from our account. We took on another employee at this time, his name was George, a lad straight from school, fifteen years old, very willing and eager to learn. He dug out the floor, and helped to cast the concrete base for the installation of our press. After learning how to set the tools and operate the machine we were well on the way again. There were teething troubles, many of them, and tempers were very often frayed, but eventually we won the day. The locks, which we had tooled up, were now being assembled and were dispatched in grosses rather than dozens, Dennis Withers was delighted, and it was of great satisfaction to ourselves. Father never gave me any credit for the work I had done, neither did I expect any, but when his cronies came to see him, the old locksmiths, some of them in the eighties, they would sit and talk about experiences in their past, I used to think they were old codgers, but in later years I realised just how clever they were. Anyway if at the time things were going OK with us, father would say to them, "don't yer think we've done well?" But if we were experiencing difficulties, the remark would be, "I'll atter watch im or ee'l ruin this bloody business."

I remember, always in the heat of an argument, our voices would be raised; a broad Wolverhampton lingo would fill the workshop until it became quite a comedy to listen to.

In later years I have realised that my father appreciated me more than I ever knew, but dialogue between us was very poor, and a barrier existed which was never to be removed. However we paddled on slowly, my transport was on its last legs, the little old 1933 ex post office van was ready for scrap, the last big journey she had made was to Sheffield and back home in a day. It was a round trip of 160 miles, on the way back the brakes failed at Ashbourne, down the big hill she went, hell for leather, round the bend at the bottom and away. At the same time I noticed a flash, the starter switch had shorted against the floor. A petrol drip from the tank above caught fire and we were having a flame up. So I smothered it with an old cloth until the van rumbled to a standstill.

It was just too much for the old girl, her brakes were operated by cables running round brass pulley wheels, the petrol tank was just in front of the dash, and gravity fed, she was out of date and soon laid to rest in the old scrap yard. So I was without transport for some months. It was now early 1955, work was on the increase, and we had a few more small customers on our books; it was time for us to employ another locksmith to make more of our hand made locks, for which there was still a heavy demand. However good locksmiths did not come easy at this time, many had left the trade for higher wages elsewhere. Young people were not interested because the lock trade did not offer enough money for them to learn, so they would seek employment in other industries such as building or labouring. We were lucky to have two youngsters, George and John, but it would take some years before there was any benefit from them. Our only way was to advertise, which we did, only one applicant turned up, and who was that, none other than the Union Reb. Oh dear, what were we going to do? A good locksmith, a troubleshooter, but we needed this type of craftsman, and so he was back with us again, in the fold. There wasn't a lot of trouble with him this time, although he would try to get us to join the trade union, but if he overstepped his mark the old chap would step in, because father wasn't paying full rate anyway.

However, I remember a spot of bother on one occasion with the Reb. "Why doe yer paint the workshop, and mek it look tidy?" he said. "Ya'll 'ave the factory inspector round and then ya'll atta do it." He was right. We did have the factory inspector, and we did atta do it. This was where

I made a real bloomer. I bought some paint on the cheap, aluminium paint, and sprayed the whole place, except for the office. I then switched on the power. By God, what had I done! Like a fool I had sprayed into the power sockets! Blue lights everywhere! Blackpool was nothing compared with this! Anyway a few weeks later the factory inspector returned and gave it the OK. He then noticed a green lead was not connected to one of the machines. "Are well" said the chap, "that's only an earth wire, it's not doing anything," he was quite good at pleading ignorance when things weren't going right. "Don't you know that it is the most important wire of all?" said the inspector abruptly, and insisted that it be reconnected immediately.

During this financial year, 1955-56, we were a little better off than in the past years so I thought I would approach the old fella about the possibility of me buying a car through the business. It was agreed that he would trade in his Ford Consul for two Ford Prefects, which of course had to be bought under finance. I was surprised when he agreed to do this because I knew his feelings about borrowing money. Anyway we purchased two new cars, payments over two years. Although we had made a little more profit this year, there were problems with Dennis Withers over price increases for our locks. Our loan to Samuel Withers was now repaid, and they were well stocked with the locks which we had put onto the assembly line. They were always pleased to take more at the price, which had not altered from six shillings and three half pence each, three pounds thirteen and sixpence per dozen, so if we were to make more profit we had to up our prices. I went to see Dennis, he was adamant when talking money. "More money," he shouted, "I thought they would get cheaper now you have got down to production. I'm not paying any more money and that's that," he shouted as he stuffed his tobacco into his pipe, and chewed irritably at the stem. Then he fumbled a match from its box, and sucked away until it was well alight. He flicked the match into the wastepaper basket behind him; I sat and watched the flames rise from it. "Why didn't you tell me the bloody basket was on fire!" he bellowed, as he choked, "get out, go on, opp it, and don't come back." So I went back to the factory to tell the old chap the news. I was in trouble again. I had upset Dennis Withers, father's favourite customer. Had it been left to me I would have finished with him altogether, he was a tyrant, but he had another side to him that one could not dislike. However, later that day Dennis telephoned, he knew that he needed our locks just as much as we needed our price increase. So he offered an extra three pence per lock. This was not acceptable in the long term, but we had to go along with him, and look out for a better market in the near future if we were going to make any money which would enable our business to run comfortably. So for the next few months I made samples of all our locks and fitted them with Perspex tops that enabled them to be inspected at a glance. I had a professional display case made into which I mounted my samples. I was ready to meet anyone. But where was I going to start? I knew the places to visit; I also knew that if Dennis Withers found out I would again be in trouble with the old fella. Although he had approved of my intentions he would not have been very happy if we had lost the custom of Samuel Withers, to whom he was always grateful, because they had given him his first orders at the start of his business during 1946.

So it was decided between us that I went to London. I made an appointment with the Crown Agents to the Colonies at No. 4 Millbank. The interview was a great success; I had impressed their inspectors with my samples, and answered all questions concerning the locks. Also I had provided accurate details of the working of our small factory and its layout. "You will be hearing from us in due course" they said. I left feeling very proud of myself, and felt that I had achieved what I had previously thought to be impossible.

However it was inevitable that there would be repercussions from Withers because they had bought our locks priced at six shillings and three half pence each, sold them to South Africa through the Crown Agents for as much as one pound fifteen shillings each and that was the main reason for my visit to the Crown Agents.

At the end of my first day after having made quite a number of other calls, I spent the evening sight seeing. I was on a tight budget and ended up with a fish and chip supper and I slept in my car overnight to save money. The next morning I went for a wash, change and brush up and continued with my business in the city for the rest of the day, again to my satisfaction. I had made many more contacts and there were good prospects of future orders, but whilst driving home in the evening I felt uneasy, something worried me.

Before I left the offices of the Crown Agents I was asked for the name of a substantial company with whom we were trading, this was for reference purposes. There was only one to fit their request; it had to be Samuel Withers and Co. Ltd. Although I knew of the possible complications which could and did follow I was not worried at the time of my interview, because I was so overwhelmed by the success of it. I also knew that Dennis Withers was well in here, and I was not poaching on his territory because he was a maker of safes and strong rooms and we were lock makers, but he could be a tyrant as we had already found out, anyway I continued my journey homeward in a jubilant mood, it was too late for regrets. When I went back to work I gave the old fella the good news first, he was real chuffed, then I told him about the reference, he blew his top. "Oh you've gone and upset him av yer" he said, "I can't leave yer to do anything properly." "Well Dad," I said, "we can't do anything about it now, we'll have to wait and see what happens." And we did, and it did happen. A week or two later, Dennis was on the telephone ranting and raving. "What do you think you are doing? I've got a letter here from the Crown Agents asking for a reference, you'll get no bloody reference off me, you've gone and bitten the hand that feeds you." "No Mr. Withers," I said, "the hand that feeds us is empty." "So that's what you think is it? Well you won't have anymore work from me if you don't drop this bloody nonsense." "Sorry Mr. Withers, but we have got to look after ourselves in the future more than we have done in the past or we'll never get anywhere." He seemed to back down and said, "OK then, what are you doing tomorrow?" "Working" I said, "why do you want to know? "Will you meet me at the Victoria Hotel in Wolverhampton, and we can talk over lunch?" "OK, I'll see you there at midday" I said. So we met as arranged. He bought lunch, some drinks and we talked, but neither of us would give ground, but I couldn't afford to anyway. He reminded me that he had friends in high places. I felt I was being pressurised, and I left his company abruptly. It was not long after our meeting when we received a letter. I opened it. The envelope was marked the Crown Agents to the Colonies, No. 4 Millbank, London. I was thrilled until I read the contents which were: 'sorry we are unable to place you on our list at this time, reasons for not listing your company will not be given,' or words to that effect. It was a bad setback for me at the time, I felt stunned. However we still had our work from Withers & Company in spite of the threats Dennis Withers had made. But I felt sure that he had used his friends in high places, which could have been the reason why we were not listed. There were other enquiries resulting from my London visit and many small orders were now being placed with us.

Another breakthrough came at this time from the Western Transvaal, South Africa, with orders for best quality hand made safe locks, payment arranged by letter of credit, through Barclays D.C.O. in London. There were repeat orders for some years and they were very valuable to us. However, just after we had received our first order, Union Reb gave notice, he was leaving again, so we employed another locksmith. He was quite a card, a practical joker and seriously claimed to be a descendant from a court jester to Queen Elizabeth I. However, he was to make the first five hundred locks for our export order. I remember the old chap saying, "how are you getting on with the locks Bill?" "OK gaffa, everyone's a cock, can't ya 'ear um crowin" he said, as he turned a key rapidly within one of the locks. However, when they were finished it was my job to pass them out and I found that he had loosened the cap screws, which enabled them to work freely. So it was an all night session for me to put them right, by thinning the top lever on each lock and screwing the caps up tightly. The next morning Bill asked, "were me locks OK gaffa?" "OK be damned" said the old fella, "none of them was cocks, none of them was crowin, they was dead, and ees ad ta stay all night ta put the bloody things right." And that was the end of the court jester.

Chapter 3

LOCK IT UP!

I had had my new car about six months and I was very proud of it, but one day I parked it outside my accountants' office. I left the keys in the ignition and the radio was playing. When I returned five minutes later the car had gone. It was quite an embarrassment, especially for a locksmith. I reluctantly went into the police station feeling a right Charlie, and reported the theft. They were clearly amused when I told them I was in business as a security locksmith. I left it in their capable hands and after three days the car was traced on the border of North Wales and Shropshire. The driver was apprehended and taken to Oswestry police station. I was asked to go there with two detectives so that I could check and identify my belongings and bring back my car.

Amongst the contents there should have been a new pair of trousers, they were missing, so I informed one of the detectives. "Come with me" he said. I followed him along a corridor. "Look through that peep hole, are those your trousers?" he said. I looked, it was the first time I had ever seen the inside of a police cell, little did I know it was not going to be the last. "Yes, they are my trousers," I said, as I watched a young man pacing up and down on six inches of flannel overlap! The door was opened and the prisoner was instructed to remove the trousers. "What is he going to travel in?" I asked. "His underpants if he has got any," said the detective. I insisted he was not to be debagged and gave him the trousers, which made me very unpopular on that day. Later at the courts I met his father, and I learned from him that his son was not a criminal, but that he had suffered mental disorders. At the trial he was found guilty and was given a suspended sentence.

Chapter 4

BACK TO THE IN-LAWS

After living at the old ladies flat for almost two years Ann became pregnant, it was to be our first and only child. We were on the move again, the flat was no longer suitable for us and buying a house was out of the question, because we had no personal money. So we returned to live with my in-laws. They were delighted, especially when our baby daughter, Carol, was born. We lived there for eight and a half years. During the early part of our stay there business was progressing slowly; export orders for high quality hand made safe locks were being well paid for and the demand for these was increasing all the time.

Our export customers had risen to three, but the profits from these were subsidising the lower paid work which we were getting from Samuel Withers. Also the assembly line which we had created was under pressure, obviously we had not kept up to date with our prices. In the past father had made small quantities of Post Office locks for Samuel Withers which they in turn exported to Cape Town, South Africa, under tender through the Crown Agents to the Colonies. These locks had previously been hand made by us and were under priced although Withers had made handsome profits from them. During the early part of 1957 tenders were sent out from the Crown Agents for 19,000 of these locks. Samuel Withers were invited to tender, and were successful in obtaining that order. Our quotation to make these for the Withers Company was one pound two shillings and sixpence each lock. We had put in a good price, and the order was firmly placed with us.

I was delighted. It was an order worth over twenty thousand pounds, just what we needed. This was going to be a great boost for the business. We would need to employ at least two more women to operate the presses and other machinery Although it was a sub-contract job to the Crown Agents, I considered it to be an important one, because some of the existing press tools could have been a valuable asset, and we had a large amount of extruded, and sheet brass in stock. But it was not to be. The final blow came when the old chap went to West Bromwich to see Dennis Withers. He returned in the afternoon in a very jubilant mood, "I've done a bit of good business today, look at this!" he said as he waved a cheque in front of me. It was from the Withers Company, a cheque for seventy nine pounds. "What's it for?" I asked "they don't owe us any money." Well ya know that order we've got for them Post Office Locks, Dennis doesn't think we could make them in the requested time, neither do I," said father, "so I've sold em back, and he's going to order them from Lowe and Fletcher of Willenhall." "You've done what!" I shouted. It was unbelievable! Was I hearing things? An order worth twenty thousand quid gone, chucked away down the drain! "Why?" I stormed, "why didn't you discuss this with me?" He didn't answer. I was dumbstruck. Had I heard him correctly? Yes I had. He was paid nineteen thousand pennies to relinquish the order. Seventy-nine pounds, sold for a song.

Dennis knew what he was doing, he found a cheaper price, and the old man had fell for it. However there was nothing more to be done except grin and bear it. I am sure he regretted his action, because there were repeat orders for these locks over the years. They were used for ballot boxes in South Africa, safes and wall safes for the home market. But we never had another enquiry for them. I was very upset for a long time; my future was worth seventy-nine quid.

I had no alternative but to carry on. Our business was priority number one for me; I had put all my effort and money into it. Father and I were daggers drawn; although we stayed together for a few years it was the beginning of the end of our business relationship, which was never very close. I could forgive him for what he had done, but I could not forget, especially at times when future orders came onto the open market for which we were not asked to quote. Everyone had a chance in life, I had two chances and now they were gone.

For some unknown reason I always felt sorry when things had gone wrong between us. Maybe it was because I knew deep down that his actions were not intended to hurt, but to protect and keep a hold on what he had worked so hard for all his life. As a father he was a good man, but to work for or to work with, he was not.

Mother was a very house-proud person, and wouldn't let him into a best room unless the furniture was covered; Covers were only removed at weekends, or when relatives came to stay. He had to take off his shoes before coming into the dining room. Perhaps this was part of our problem. Also because I was spoilt and pampered by her. I remember a fella saying to him one day, "Bill, I cured my missus from that very same thing." "Oh, what did ya do?" said the old chap. "I took my shoes off and threw them through the winda and it worked." "I might try that," said the old fella.

However there was a night when he threatened to do this, but he didn't carry out his intended action, maybe it was because he was scared, or considered the repair work he would have on his hands afterwards.

There were many times when he was not well, and mother would say, "Come on shake your feathers and get off to work." "Ah, you wait til the black crow treads on ya" he would say and off to work he'd go as instructed. I often wonder where the saying about the black crow came from.

It was now late 1957 and at this time my brother Geoff approached the old chap for the possible use of part of our factory and to be able to use the power press and hand presses.

Geoff was a draughtsman by trade. He had been contacted by a chap in London who had asked him to design and make press tools for the production of large quantities of household kitchen gadgets, which he wanted to produce on our premises using female labour. The utensils or gadgets were to be sold at Ideal Home Exhibitions. The old fella gave it the OK, and our kid worked hard to get the line moving. His customer was a dapper little bloke, a one-man business, with a demonstrator. She was a smart bit of skirt who looked after his affairs both private and business, in and out of bed. I bring this into my life story as a locksmith because it plays a significant part later.

Chapter 5

FIFTY QUID FOR A KEY

I remember it was a day when the old fella was cutting a lot of safe keys to pattern. Each pattern key carried a label bearing the name and address of the sender, and a description of the safe's usage. When our factory was closed all these keys were locked in a safe, and were ready to be dispatched the next day. On this particular occasion a fella called in to ask if we would cut some keys and repair a lock for him. He was an odd job builder and had been in many times before. "Bill, can you oblige me by doing this job whilst I wait for it?" he asked. "Yes, OK," said the old chap. So he waited, and watched the work being done.

Then came a bolt out of the blue. "Do you want to earn some quick dough?" he asked. "Don't we all!" said the old fella. "Well" he said, as he flashed his beady eyes across the labelled keys, "you've got a safe key there marked wages safe, George Mason's Grocers, I'll give yer fifty quid if yer cut me one to pattern, and I'll be in and gone with the dough before morning." "Ah," said the old chap, "and who are the police going to suspect? Who do you think I am? Billy Muggins? Out, before I kick you out." "I'm only joking" he replied, but he wasn't because a week later we heard he was caught whilst escaping down a drainpipe after breaking and entering into houses nearby. Anyway he went down for a spell, and that was the last we saw of him. After this incident all our customers' keys were coded.

Chapter 6

GENTLEMEN IN BUSINESS

For some considerable time I had been practising the art of manipulating locks by picking, and many other methods. This was to be another string to our bow, a skill that the old fella had not ventured into. It had prospects, so I sold my car and bought a Morris Oxford Brake, and equipped it with a workbench at the rear. I could work at the bench with both back doors wide open, and I made a detachable roof that enabled me to cope in all kinds of weather. I mounted a vice and anvil on the bench, and packed the car with tools that were vital to my trade. It was March 1958 when I first put my acquired skills to the test. On the back doors of the brake our name was printed: -

WBS Safe Locks Ltd
Safe and Strong Room Engineers
Telephone: - Wolverhampton 31624

Emergency calls came by telephone, and within a few minutes I would be on my way, but I was soon to find out how little knowledge I had.

I always studied each job closely before and after opening, made notes, and learned from the many mistakes I made. At the start I lacked confidence and needed much more experience. This came automatically as the work built up. For a while I worried about failure, because when you are faced with a strong room or safe door, (locked), and you are on your own, believe me you have to think all the way. I will never forget the first safe I opened. The feeling was great, and I had conquered it. Any locksmith involved in this type of work will always remember his first triumph.

I never failed to open a lock whether it was on a house door, safe or strong room doors, but I do know that today's equipment is so sophisticated, even the best could fail, and I didn't claim to be the best.

In the early days one had to be successful to justify payment from the customer.

At this time we were being paid between five and seven shillings to cut conventional safe keys to pattern for locks ranging from six to eight levers, little more than the price of a small can of beer today. Cutting an eight-lever key from a blank by hand would take me about forty minutes to an hour, however this was quicker than returning to the factory to cut the keys by machine. I could also charge extra, and the immediate service was greatly appreciated by the customer.

This was only the beginning of my career as a locksmith working an emergency call-out system, so I worked locally knowing that I had the factory as a back-up, although it was not very long before I was to take on more work in many parts of the country.

I made contact with a firm in Birmingham. They were lock makers and manufacturers of cell doors, grille gates and all types of equipment used in police stations. I was soon to become their service engineer. It was a pleasure to carry out work on their behalf; I had for the first time met gentlemen in business.

The offices and factory were in MacDonald Street, Birmingham, and their family home was in Selly Oak. I must tell you something about the family, because to me they were unique.

They were a family who managed their affairs without aggravation amongst themselves. There was the father, the mother two sons and a daughter. All took part in the business, each having an important role to play, and each member of the family left home for business in their own mode of transport. The older son was married, he lived next door to the family home, and travelled in his Austin Healey sports car to the factory each day. The father was over eighty at this time and went to business in his Daimler, the mother travelled by bus, the daughter went on her bike and the younger son went in his MG sports car.

A total of five employees worked on their factory floor. Whenever they had work for me, I was asked for an approximate quotation, not like other firms I had dealt with in the past, who had tried to dictate the price they would pay, and were very often successful.

I was very happy working for this firm, and we became good friends. Business was conducted in a gentlemanly way. I remember being invited to their home one Sunday afternoon. It was a beautiful hot summer's day. We were to play croquet. Before we arrived at the house we were quite apprehensive, as it was a very old-fashioned game to play. However it turned out to be great fun, there was an interval during the game, fresh cream cakes were served, and we were told that this was traditional before resuming our play. Later we went indoors and were given a very enjoyable evening meal. Our business relationship was to continue until their business was sold and the family went into retirement.

Chapter 7

WORKING INSIDE

I was quite happy as our outwork increased, and being out of the old chap's way was good for both of us. He looked after the factory and kept our few employees in check.

My first job working inside the cells was very different from what I had been doing, another job to remember; it was like opening my first safe. Ann came with me, we were travelling to North Yorkshire. I always remember this one because after about four hours on the road we developed car trouble. The fan belt had broken, we were miles away from anywhere, what was I going to do? It would be hours before I could get away from this desolate place, and I had an appointment at the police station where I was going to work. Suddenly I thought "Ann, take your nylons off, I need them for a fan belt." "Alright," she said "as long as that's all I've got to take off!" "We haven't got time for anything else, there's a job to be done, you'll have to wait until we get back home for that!"

So I made an improvised fan belt and off we went for a few miles. We coasted down hills with the engine switched off until finally the nylon stockings were in threads. However we were lucky, there was a garage in sight. A second hand fan belt was taken from an old car, fitted to my vehicle and we were on our way. On arrival at the police station I made my apologies and I was permitted to carry out my work, which continued well into the late hours of the night. My instructions were to check all cell doors, food hatches, gate locks, both male and female blocks, and carry out repairs where necessary.

I started work. The noise I created from hammering echoed throughout the long corridors for most of the night, and I was in trouble, not with the law, but with the prisoners. Abuse, I had it all that night, and I have heard it many times since then.

After I had completed my work on the first cell, I would report to the sergeant at the desk and he would move another prisoner out which enabled me to continue to do the job.

I remember one of the prisoners trying to wind me up whilst he was being transferred from one cell to another. I ignored him. "Aye I'm talking to thee!" he said "if thou's got a home to go to, piss off and let's get some f***in kip". Others would ask, "Give us a fag mate."

There was another fella who shouted, "Why don't you sling your hook and bugger off!" "Yes I will when I have finished," I said "but until then you'll have ta bloody well put up with me." Mind you the noise I made and the abuse from them was enough to aggravate a saint and I was no saint. Anyway after a lot of barracking during the night I finished my work, shouted "ter ra" to the prisoners who's quick response was to "f*** off" and "don't come bloody back again". However we were well fed in the police canteen and left for home in the early hours. My first job working inside was a complete success.

I continued with my service work, which was building up quite well, especially with the police stations and law courts. The old chap carried on at the factory. Brother Geoff got more involved with household gadgets and kitchen utensils, which seemed to bring in a lot of debt through bad payers and twisters.

Chapter 8

BACK AT THE FACTORY

At the factory we were still making locks for Samuel Withers. Needless to say the price of our locks which I had developed on the assembly line were again falling behind and we were under pressure due to rises in living costs.

Wages day seemed to come round so quickly, even though I had to work seven days a week, and all hours. We were always short of the ready and I would obtain cash by changing cheques with the milkman, sometimes over a hundred pounds at a time, which in those days was a lot of money. This amount would pay wages for seven skilled workmen including some overtime. I would pay the wages, and work though many weekends, without going home. This enabled me to deliver our locks to Withers on Mondays. In return they would deduct their discount for prompt payment, and the cheque would be paid into my bank before the Dairy cheque was presented. It became a regular practice for some years. I must have taken thousands of pounds off the milkman during this time, his job was on the line but I never ever let him down, if only he knew the risk he was taking, then I'm sure he would have refused, but I valued his trust in me. Also I found it to be a better way than going cap in hand to the bank manager who would have bounced a cheque for as little as five pounds in those days. I was already labelled by my bank manager as the busiest fool he had ever known, "why don't you increase your prices?" he would say. This was easier said than done because we had to get our wages in whilst we were in the process of making locks for export to South Africa and the Western Transvaal. Our weekly cash flow was most important.

Competition was also very keen in the lock trade. Locksmiths worked in back street workshops, brew houses and sheds in the evenings after being in regular employment during the daytime. It was cheap labour and in most cases was for beer money. This was quite a considerable contribution towards competitiveness in the business, and the lock trade in general. There were plenty of employee locksmiths who had no sense of responsibility, and drinking was their social way of life. Many of them were unskilled. I remember one of my workmen, Harry was his name, and he was a good grafter but not a great locksmith. Also he was the most unreliable character that I have ever met. He owed money all over the place did Harry, and had unpaid hire purchase agreements everywhere. There are three brief tales I can tell you about him. But firstly a bit about his home life. He had a few snotty nosed kids, a little Irish wife who was frightened of him; she couldn't bake or roast, so she boiled everything from a pudding to a pair of socks. Now 'H' always had a telly, even though the place was sparse, and lacked furniture, although there were a couple of easy chairs and a bench seat in the front room. Harry was a likeable bloke, and his door was open to any and every one of his mates who wanted to call. My first tale about him took place during the Tennis Tournaments. It was an afternoon when he was about two hours late coming back from his lunch, when I asked, "Where the hell have you been Harry?" "Well" he said, "I 'ad a bit of trouble, I was watchin Wimbledon on the box, when somebody banged on the door. It was open so I called 'cum in, I'm in 'ere watching the tennis'. So in cums two blokes, and one bloke said 'we are from the television licensing authority, do you have a current licence?' 'No mate' I said, 'I've got a good enough picture without one.'" So Harry was fined, but as usual I had to pay it for him, and enter it in the sub book, which was another ten bob a week and missings. I remember one particular time when Harry was really on his uppers, although it was hard to remember a day when he wasn't. Like most drinkers he was always unhappy when he hadn't got any money for his beer, and this was one of those frequent occasions, so he came to me, as was usually the case and this tells my second tale about Harry. "Gaffer, I need twenty quid, how's it looking," he said. "Don't we all Harry," I said, "not too good." "I'll sell you me telly for twenty quid, do yer wanna buy it?" he asked. "I'll bet it's on the never-never" I remarked. "No it ain't," he replied, "I'll show yer the receipt paid in full," Harry never paid for anything in his life, so how could he have a receipt, it was a load of old codswallop, at least I thought it was, but I was wrong, he'd got one this time. So finally I bought his telly, we agreed a price of eighteen quid. On my way home that evening I took Harry with me, and we went in his house to get it. As I have already said, Harry hadn't got much furniture; his kids sat on a long rough wooden homemade bench, and were happily viewing the box that I had come to collect. So Harry pulls the plug, off went the set. "The masters cum ter tek the telly," he said, and then the bartin began, there was uproar, and the house became a bedlam. "Oh put it back on Harry," I said, "I'll pick it up tomorrow." "No tek it off, tek it now, I'll have a new un in before the wick-end, and it'll be a coloured un." So I took it, and he did have a new one, and it was coloured. Over the next twelve months Harry substantially increased his sub book account to pay for it, and once again I was Billy Muggins.

It seemed funny that Harry could have a new colour telly, sub money off me to pay for it, and I bought his old black and white one.

The last tale about Harry was that one-day whilst he was working for me on an export order there came a knock at the office door. My secretary answered it and called "there are two gentleman here to see you." "OK Maureen, I'll be with you in a minute." So I went to the office. "Good morning" I said, "what can I do for you?" "We understand that a Mr. Harry B is in your employment." "Yes he is," I answered. "May we please see him?" "I am sorry" I replied, "but we have a very important export order in progress which has to be at Southampton Docks, and he is working on it at the moment. Can you come back later or see him at his home?" I asked. "Let me explain," said one of the gentlemen, "we are committal officers from Winson Green Prison and we do have a warrant for the arrest of Mr. B." "Well" I said, " I'll fetch him and see if we can sort this out." He had a debt of nine pounds plus costs with Raleigh Cycles, so the only way I could keep him out of jail was to pay the fine. This was another ten shillings a week to go on his sub book. Harry was not a criminal, but he just could not accept the responsibilities of life. He thought everything should be free of charge especially to him. However, he worked for me for some years, and his sub book was always well stocked with loans. I remember he would regularly comment, "you will never be as well off as me Gaffer, because I've got nothing, the state will keep me." And they did for a good many years. Eventually he got on the disabled list and the last I heard of Harry was that he was doing well on the Social Security.

Chapter 9

KITCHEN GADGETS

My brother Geoff had been making these kitchen gadgets for sometime now, but things were getting increasingly difficult for him, because his customer was not paying the full amount of his bill although he was crying out continually for more stocks.

Materials were very costly, and had to be paid for immediately either on collection or delivery from the manufacturer. This meant that a cash flow was most essential to enable him to survive. Finally Geoff packed up the business; hence his customer approached the old chap and me, and asked if we would take on the production for him. We thought about this very carefully because there was a great deal of mistrust between demonstrators, stand holders, and manufacturers. However, it was decided that we give it a go, but only on one condition, which was to be, payment in full cash on delivery or collection for all purchases, or cheques which were to be cleared before the release of goods. So we registered with H.M. Customs and Excise for purchase tax.

The four women that Brother Geoff had employed were kept on, and became useful in lock assembly. Kitchen gadgets consisted of graters made from high quality nickel-plated steel, shredders, peelers, knives, potato scoops etc.

It was all a bit of a mushroom effort. The bloke and his 'in and out of bed' demonstrator turned out to be a couple of fly be nights in more ways than one and after twelve months the whole thing fizzled out anyway.

So it was time to cut our losses, and inform HM Customs that we wished to be de-registered from purchase tax. One of their inspectors came to visit, us I remember him very well his name was Lewis, and a right pompous arse, he was. "All these parts you have in stock for the making of kitchen utensils will have to be disposed of", he said. "Well, what about the wooden handles", I asked, I've got grosses of these which cost me fourpence each," "Burn them outside in the yard", he suggested. "No way will I burn 'em outside", I shouted, "I'll light the stove with them every morning until they've gone, and if you don't trust me, then you must bloody well come every day and watch me, and that's how it's going to be".

So he agreed, but even so he made random checks until they had been burned. Now regarding the metal parts, "what are you going to do about those?" he said. "Well Mr. Lewis", I replied, "I'm going to dig a hole outside to put them in". "Do you mind if I watch", he said, "No" I answered, if it satisfies your petty nature then join me". So I dug a large hole about five feet deep, sweat my cobblers off, I did, then I put all the parts in the hole, gave the spade to Mr. Lewis, and said "here, you fill the bloody hole in, it's my turn to watch now". That was the end of another escapade, which resulted in a money loss.

Chapter 10

CONTEMPT OF COURT

One day I was going to inspect a Strong Room Door, with a view of carrying out extensive repairs to the hinges. The job was at Lichfield and was underneath the Law Courts. After my inspection I advised the authorities that there would be considerable noise involved whilst the work was in progress. Therefore a time and date had to be arranged when there were to be no law courts in session, so I returned at the agreed time, I took all my tools down to the dungeon-like basement and proceeded to carry out the work. I hadn't been working long when a fella came downstairs and said, "you will have to stop the noise, there's a court in session", "sorry" I said, "but this work is by appointment and has to be finished today", so he went away, and I continued to hammer on, until another bloke comes down this time he was in uniform, "you will have to go away and come back another day", "who says so", I asked, "the judge", he replied, "well", I said, "here is my authorisation, go and ask him who is going to pay, so I carried on again. The third visit was from a higher rank, "if you don't pack it up you will be arrested for contempt of court, and that's an order from the judge", he said. Well who heard of a Locksmith being locked up whilst carrying out his lawful work, so I packed up went back to the factory and reluctantly, they paid me for the error that they had made. Another appointment was arranged for me and I was able to complete my work there without interruption.

Chapter 11

A FEW INCIDENTS

TEARS TO THE EYES

I remember one day when I went to work at a Birmingham Police Station I was to carry out repairs to door and gate locks in the female cell-block There were many arrests whilst I was there. I remember in particular a young woman was brought into custody, she was scantily dressed and wore a pair of stiletto heel shoes which she used in a vicious attempt to free herself by kicking the two officers on the shins, until eventually she was brought under control, her shoes removed, put in the cell and the door secured. After closing the door one of the officers noticed spots of blood on the corridor floor, she had obviously suffered some injury. Following this a senior police officer was called in to check her out, possibly because she had gone berserk anyway. On his arrival to the cell the door was opened by a custodian officer and like a bolt of lightning she flew across the cell and kicked the officer in his goolies. He was right in the line of fire, and ended up on his back against the corridor wall. He was flat out and obviously in a lot of pain. Well it brought tears to my eyes and I was only watching.

HOW DID THEY GET THERE

A strange incident happened at another police station I visited in the North Country, where I had to inspect and carry out any necessary repairs to the locking mechanism and food hatches.

It was a male cell-block. There were eight male cells in all and the prisoners were moved accordingly so that I could do my work. One of the prisoners detained there was a well-known professional burglar. During the early morning his breakfast was taken to him and it was noticed that there were footprints on the ceiling of his cell, nobody actually knew how they had got there. The only possible theory, was that he may have somehow used his bedding for a head stand in order to be able to reach the ceiling with his bare feet, but there was no conclusion to the incident whilst I was there.

WHERE DID HE GET THE STRING

Reports came through to us stating that a prisoner had committed suicide by hanging himself with a piece of string anchored to the top hinge on his cell door. This resulted in a lot of work for us.

We travelled to police stations in many parts of the country, where we were instructed to weld slip cones onto the top door hinges and to test each one by pulling on a piece of string before leaving. Providing the efficiency of the custody officers was impeccable and the prisoners searched thoroughly, there should have been no string available to him. So where did they get the string.........?

Chapter 12

A GOOD OFFER

During 1960 I received a letter from two Dutch African businessmen whom I had met briefly earlier during that year.

The content of the letter was that they were revisiting Britain again shortly, and wished to contact me upon their arrival. This they did and so I made arrangements to meet them at a hotel in Worcester.

The usual introductions to their associates took place. After a few beers and an excellent meal they began to tell me the purpose of their visit. Their christian names were John and Philip, John was the senior, so he was to start the ball rolling in the way of conversation. "Well now Bob", he said, "what do you think about coming to work in South Africa"? "South Africa", I said, "I've never given it a thought". "Well, Philip and I have a proposition which we think is going to be good for us all. As you know", he continued, "we have our safe manufacturing business in Cape Town, and we are using your locks for our safes, but we need our own lock factory, this is where you come in". "Carry on", I said, "I'm listening". So he went on. "Close your factory down, sell the premises, ship out everything you need, tools etc., and all we want from then on is your expertise". "OK, but what's in it for me", I asked. So he replied, "First of all, you and your family come out to us for a holiday, stay as long as you like, be our guests, all expenses paid, you don't need a penny" he said, "and then, he continued, if you like it, and wish to join us we will form a limited company of which you, Philip and myself, will be directors, we will also build a house gratuitously for you". I promptly replied, "it's a damned good offer, but I must discuss it with Ann, my wife and consider it from a family point of view. They fully agreed, and commented, "take all the time you need, then come and have a holiday with us". So we shook hands and continued with an enjoyable social evening before we parted.

After much discussion Ann and I mutually agreed that to leave our families and our country was not something we wished to do. As it happened we felt that we had made the right decision, because not long after this meeting, trouble in Rhodesia was building up under the leadership of Ian Smith.

Chapter 13

BACK TO SCHOOL

We had repeat orders for locks for export to Cape Town and the Western Transvaal. Also the home market, which was always flagging behind regarding prices, was improving. I was coping OK with the outside work and emergency call-outs were frequent.

Young George was becoming very useful now. I needed to keep him busy, so I approached the education authorities, and asked if we could place ourselves available to carry out lock repairs, and maintenance on municipal buildings. Most of the work we were contracted to do was for primary and secondary schools. These were within about an eight mile radius of our business, and there were a lot of schools.

It was a job which carried abuse with it, this time from the kids.

When we worked on school doors adjacent to the playground it was like being back at school. Invariably the kids were outside for playtime. They'd throw pebbles at us, call us all the names they could think of, and there was nothing we could do about it

One young lad said to me one day, "what ya doin mister?", "I'm, mending the locks", I replied. "Well you mend em Gaffa and we'll bust em", was another comment.

I remember one little horror, a real scruffy urchin with steel rim glasses, one eyeglass blanked off. He was playing football in the playground and kicked the ball through the window. When I went back the next day to re glaze it, he was there again, and he pointed to another window, and said, "oh look mister there's another up there broke". We carried on doing this work for a few months and finally decided that our schooling days were over.

Chapter 14

A WORKS TRIP TO ASCOT

It was 1961 Ascot Gold Cup day when I had made arrangements to take the workers on an annual day outing. It wasn't my type of outing but it was their choice. So I hired a mini coach, self drive, and we were off for a day at the races. There were about eleven of us in all including Ann, my wife. We pushed off at about six o'clock in the morning from my in-laws house where we lived, called for the old biddies that worked for me: they all had their big hats on and they piled into the coach. Then we went to pick up the blokes, who were dressed in their best suits and off we went.

Taking the scenic route and enjoying the countryside, we arrived at Ascot at about ten o'clock in the morning. After a brief look around, we came across a spot ideal for us to have our picnic lunch, and they talked about how much money they were going to win. I wasn't interested in betting anymore but Ann joined in and, like most people who go to the races, they lost on the day. In fact the women in our party were taken for a ride literally, when a photographer took pictures of them, took their money and their names and addresses, and that was the last they ever heard of that.

After a good day out we started for home, and on our way back we pulled in at the hotel for our evening meal, which I thought would make a nice finish to the outing. After the meal we made tracks towards home, when there was a loud bang, the back tyre had burst. "Won't be ten minutes", I said, "and we will be off again". When I looked in the back of the coach there was a spare tyre, but no jack or wheel brace. It was getting late about eleven o'clock at night, but we were lucky enough to get help from a nearby garage. This wasn't the end of our troubles, because we hadn't gone above twenty miles further before the engine started playing up until finally it conked out. "Well there's only one thing to do, we'll have to push the thing to the top of the hill", I said. So with big hats and best suits it was down to the nitty gritty and shove.

The women were in high spirits, whilst the men did all the grumping. As we coasted down the hill I saw a light in the distance, so I pulled up. We were in luck again, there was a small haulage business and men were

working all night and they got us out of trouble and once again we continued on our journey. We arrived home at about three o'clock in the morning. No one turned in for work until the afternoon. We never had another works' outing.

Chapter 15

THE "REBEL" RETURNS

It was during 1962 when I couldn't help but notice that the old fella was rapidly losing interest in the business, he was almost sixty five, and he had got another job lined up in Wolverhampton. He was a master craftsman and firms were only to pleased to employ a man of his calibre even though he was approaching retirement age. Also he would be free from responsibility, which was not one of his finer points anyway. There were hints that he would like to take his money and pull out. So it was decided between us that he should be paid an agreed amount of money, also take a car, and the business should continue to pay his house mortgage, which was about twelve pounds a month until completion of the amount agreed for his share. If he didn't survive, then monies owing would automatically be paid to mother.

It was also agreed that he would not leave until I could find another locksmith with similar capabilities to himself, but this was going to be impossible. Craftsmen were becoming increasingly more difficult to find. So I advertised continuously. There was only one applicant during a period of almost six months and he was the last person I ever wanted to see, that bloody Rebel. "What do you want", I asked, as if I didn't know, when he knocked on my office door. "Arve cum about the job, I eer ya want a foreman", he said, "and I'm ya mon". I was seething inside, how he had the nerve to come again for a job was beyond my apprehension. "You'd better come in", I said, "and I'll decide whether your me mon or not". He seemed more docile on this occasion, well he was getting older, and perhaps he was looking for an easier job, also a job as a foreman would be good for his ego.

There was very little I could do other than set him on as foreman, so I gave him the job, although he wasn't good enough to lace the old man's boots up, but he was the best I was going to get. His salary was agreed, at twenty-four pounds a week and he was going to look after the twelve employees whilst I continued with the outside work.

I had to do things a bit different than I had done in the past. In other words be stricter regarding time-keeping, lunch-breaks and dinner-hours. So the first step was to buy a time-clock. Well, it's not easy to install a time clock, especially when for years, everyone's come in willy nilly. There were remarks yes, plenty of 'em, and as I expected the Reb was the star of the show, and up to his old tricks again. For instance, I would arrive at the factory somewhere around eight thirty most mornings, and I was regularly greeted by his uncouth remarks: "we atta be ere at eight in a morning, cor ya gerrup", he'd say. His attitude had been familiar to me in the past, so I chose to ignore him, but after some time it began to rankle. Well I was the gaffer and I was paying high rates for his work and his insults. So I decided to play him at his own game. One morning I came into work early, before anyone had arrived. "Well, what's the matter", he asked, "you'm early aye yer, wot ya dun, s**t the bed". This was a situation that had to be ignored, the old proverb was true, the leopard never changes his spots.

At this time the old fella had left and started his new job. He wasn't very happy about the Reb being there. He would come and see me and we would go for a beer together, in fact we got on with each other much better than we had done in the past.

About this time Dennis Withers had suffered a fatal heart attack and within a short time he was dead. It wasn't long before the Withers Group came under new management and Dennis's wife had taken the chair, she was also a director of the company. The new management was not very efficient and soon the tide was to turn in my favour. A short time after this I was contracted by the Withers Company and asked if I would take on their safe removals, the installation of strong room doors and frames, and in particular the opening of safes. So I mustered up a crew, and we started off. After working together for some weeks we became a good team, although most of the men employed on safe removals were part time workmen. We moved safes upstairs, downstairs, spiral staircases, anywhere and any weight, five hundred weights, half ton, one ton or more. For the first time I was able to price correctly and make a decent profit. What a great pity that someone has to die to enable a change to be brought about.

It was now time for me to make a change, especially in the direction of transport regarding commercial vehicles. So we purchased a Bedford T.K. Lorry, a Mini van, and two Italian Auto Bianchi vans. Our Lorry was a great asset, we hadn't had it more than a few weeks when I received a call from Withers requesting me to collect from their factory one vestibule type strong room door and frame of which the estimated weight was about twenty eight hundred weights. I was to install it at a new branch of the Birmingham Municipal Bank and the site was at Hockley. The building however was not completed. I was to meet a Mr. Maggs there at an appointed time, which had been made for me. Now I had met Mr. Maggs on numerous occasions and we were not on the best of terms with each other because I had always let him down by being late, unwell or having vehicle breakdowns. So I wasn't very popular with Maggs and Maggs was one of the top brass at the bank, also he was a very good engineer.

Anyway it was important to me that I got everything right this time and that there were no snags with Maggs. It was the first installation of this kind that I had ever undertaken and if I looked after it there would be more work. So three of us set out in the lorry for West Bromwich, we picked up the strong room door and frame and continued to Hockley. When I arrived there I tried to get as near to the unfinished building as possible so I ploughed across the muddy site until finally I came to rest axle deep in a quagmire. I was there on time, but it wasn't good enough, I had failed again. Along came Maggs with his oppo, "Oh god it's him, it's that Sidbotham again. What the bloody hell have you done now, there are plenty of materials on site which you could have used to get your lorry across without getting in this mess", he said. "Well er yes I'm sorry Mr. Maggs", I replied. "OK" he said, "but I'm not staying here all day to suit you, I've got more important things to do, they are your problems and you must get out of them as best you can. However I'll show you where to put the door and frame and then you can meet me tomorrow and I'll supervise the installation", and off they went. We hadn't any Wellingtons and we were knee deep in quag but after some hours of hard graft we managed to sort it out.

The next day I returned I was not very confident and Maggs knew it. "Have you ever installed one of these before"? he asked. I hesitated to answer him. "No" he snapped "you haven't have you"? "No sir", I replied. "Then I am going to show you how to build in a strong room door and frame so let's get on with it", he said, and the job was done. From then onwards, Maggs and I got along OK purely because, I admitted that I was not familiar with the job I was about to undertake.

Our Mini van, which cost about two hundred and sixty pounds, was brand new and it was used for delivery of small safes etc. I remember one day when I had to go to Doncaster to deliver and install a safe, which had been reconditioned at my factory. I was also to collect and bring back a safe, which was to be renovated and put into stock.

The job was for Burton Tailoring, (Montague Burton) Head Office at Doncaster. Both they and the fifty-shilling tailors were the most popular tailors of that time and had branches in many parts of the country in which I was involved. Anyway I went to my factory on that morning at five o'clock. On my own I manhandled the safe into the Mini van, its weight was almost five hundred weights and then I was on my way to Doncaster. I arrived at my destination at just turned nine o'clock in the morning. I off- loaded the safe and with the aid of a sack truck I installed it in the required position. Then, once again, I man-handled the other safe into the van, which was of a similar weight to the one I had off loaded. Before leaving for home, I telephoned my office; it was about ten thirty that morning. "Hello, Maureen, Bob here, have you any messages for me", "Yes, hold on", she said, "An emergency has come in, I've got a job at Llanelli, in South Wales, and it's a safe to be opened at a branch of George Masons, it must be done today", she added. "Isn't there anyone you can send from there", I asked. "No" she replied, "I've had to send George on another big job". "OK leave it with me, I'll go from here, can you give me the details, and then telephone the branch to say I'll be there as soon as I can".

Now George Masons were a new customer of ours with whom we were able to increase the business of opening, repairing and the moving of safes. They were a chain store in groceries and provisions, which resulted in there being plenty of work for us. Their head office was in Bradford Street, Birmingham.

Anyway I left Doncaster and drove hell for leather down to Llanelli to carry out their work. It was wintertime, a cold day, with plenty of snow, but I got there OK. The five hundred weight safe I had on board had helped to hold the back down whilst I skated over the slippery roads. I always carried my tools with me and I was not short of kit for the occasion. So I proceeded with my work, it took a long time to open that safe and I had to fit a new lock, and carry out the necessary repairs. It had been a big day, and it was going to be a long night, so I handed the keys to the manager and set off for home.

I left at about eleven o'clock at night and had to face the snow and icy roads home. I hadn't been travelling long when I almost lost the van, whilst taking a bend too fast. The safe broke loose and shot from one side of the vehicle to the other. In the moonlight I thought I saw a large drop on my nearside. I stopped to secure the safe, and looked, there was a deep ravine, I had made a bad mistake but I always learned by my mistakes. So I continued on my way with the utmost care and respect for the journey that lay ahead.

It was a nightmare journey. I was short of cash, I was starving and I hadn't eaten for fifteen hours and to top it all, there wasn't enough petrol in the tank to get me home. So I pulled up at a garage, it was almost midnight, there was a fish and chip shop next to the garage. I could only afford fuel, and not food. It was hard to make the choice, especially with the smell of fish and chips next door. I paid for the fuel, and continued on my journey coasting down hills until finally I was within half a mile from home, when the fuel pump began to tick.

I managed to pull in on my local pub car park, the Merry Hill, Wolverhampton, where I ran out of fuel. I locked up the van, walked home a quarter of a mile, had an early morning supper, and flopped into bed. I had travelled five hundred miles, and worked twenty-four hours non-stop, too much for any man!

Chapter 16

TWO FINDS AND A FINDERS KEEPERS

At this time I had made some good contacts in the Jewellery Quarter of Birmingham. One customer in particular was a Mr. D of Vyse Street. This short story reflects back to a job when he telephoned me and asked, "have you got a good quality safe", "yes I have what size do you want Mr. D, and how much cover do you need". So he gave me the size and cover required. "Yes", I said, "I've got a Chubb here, and it has the protection suitable for your requirements, it's in good condition and not very old." "How much is it," he asked. "Five hundred and fifty pounds, plus installation, and its weight is over a ton". I asked, "Do you want to come and see it"? "No", he replied, "I'll take it, can you come over in the morning and we'll sort out the best place to put it". "Yes alright Mr. D, I'll see you then", and I went over as arranged. "I'd like it put in this corner, up against the wall, what do you think"? He said. "Well, it's over a ton and it's not going to come upstairs. We will have to take the window frame out and crane it through". "Also the floor will have to be substantially re-enforced to take the weight". "Do you want a quotation"? "No", he replied, "just get on with it as soon as you can". "OK we'll start tomorrow", I said, and within two days the job was completed.

"Are you quite satisfied with the job Mr. D"? I asked. "Yes absolutely", he replied. "Now before you go, I've got a safe here which is locked and I've lost the keys, will you take it out of my way, open it, make a couple of keys and will you keep it at your place until I've got room for it"? It wasn't a heavy safe, so we soon loaded it onto the lorry, Mr. D gave my crew a tip of five pounds and away we went.

When we returned to the factory we off loaded the safe and our tools, it was stored in the yard underneath a tarpaulin sheet with many other safes and lay there for about two years until one day Mr. D telephoned, "you know that safe of mine, which I asked you to do up for me", he said. "Yes Mr. D it's in the yard, sheeted up, never touched it since I brought it in". "Well", he said, "can you do it up and bring it over during next week as I've got use for it"? "Yes, OK that'll be fine, I'll ring you before I come". So the safe was brought into the workshop and I called George over to open it, make the necessary repairs and cut two new keys. He opened it after about half an hour, when suddenly he became excited, "look gaffer, look, bloody diamonds, we'm rich". They were sparklers alright, quite large and seven of them. "Wot ya gonna do with em gaffer", said George. "Give them back to Mr. D of course" I said. But when I telephoned Mr. D he could not recollect any such loss. So I went to Birmingham delivered his safe and returned his diamonds. He thanked me and said, "Come and see me if you want any jewellery, your honesty has impressed me very much". Then he showed me around his workshop where his men were working at their benches. They wore leather aprons and collected the gold dust and filings very carefully which were put into two tins, one tin for Mr. D's holidays and one for his workmen for Christmas. "Mind you a tin of gold dust would have been a good reward for me". However, I didn't expect anything, but I was soon to discover that he was putting work my way, which resulted in good businesses within the Jewellery Quarter.

ANOTHER FIND

The next of our finds again occurred in the Jewellery Quarter of Birmingham. Our crew of six workmen including myself were moving quite a few safes from the ground to the upper floors. It was a very old building and it hadn't been cleaned out properly for years. Whilst we were moving one of the larger safes from a corner wall I noticed that one of my crew had bent down as if to pick something up. He looked a bit sheepish, and I thought he may have had something in his left hand but I was not certain, so I took a chance, and bluffed him. "What have you got there? Come on let's have it", I said. "Nothing", he replied, and opened his right hand. "No, not that one, the other one". Yes I was right, there it was a gold ring inset with diamonds. "You would have kept it, wouldn't you"? I said. "Yes, and you're a bloody fool if you hand it in", he remarked. He was probably right, maybe I was a fool, but it was my business and my reputation and most of all I was honest. "I found it on the floor, behind the safe", he said. "I know where you found it, I saw you bend down and I thought you picked something up". "I bet it's been there years and who's going to know"? He asked. Anyway I handed it to the management and although it was a valuable ring, they weren't very appreciative in fact, one of the bosses told us that tea was made but to carry on working as it was too hot to drink.

I was honest and I often wondered whether honesty was the best policy.

FINDERS KEEPERS

This time it was finder's keepers.

I was called out to unlock a safe at Sandonia Bingo Hall, it was in Stafford. The safe was a fiddling little thing about eighteen inches high and was of poor quality. It took me about two minutes to open it by the method of picking the lock. After doing so I advised the customer of the quality and condition of the safe, and I also advised him that it was not suitable for the protection of cash. "Have you got a new safe I can buy from you", he asked. "No", I said, sorry but I don't stock new safes, but if you tell me what you want, I'll get one for you within a week if I can". "Something about forty two inches high, fitted with a combination lock if possible", he said. A couple of days later I managed to get a new safe from Samuel Withers. I phoned the Bingo Hall and the Manager was over the moon. Anyway I delivered it within a few days, made thirty quid, which was a good profit at that time, plus charges for delivery, and I was well paid for opening the other one. Before I left the manager asked me if I wanted to buy the old safe. "How much do you want for it", I asked. "Give us a couple of quid," he said. "OK", so I took it off his hands at that price and put it in my yard, under the tarpaulins with other safes and left it there for about six months. Later a fella came to see me, he wanted to buy a small safe for the purpose of storing explosives. The explosives were to be used for rock-blasting in his construction business for the building of new roadways. "I'll give you thirty five pounds for that one, can you supply four keys and can I collect it this afternoon" he asked. "Yes I'll have it ready for you," I said. It took about an hour to get it ready and I had only paid two pounds for it. Before he returned to pick it up I had the presence of mind to remove the drawer, and check at the back. There I found a roll of twenty ten bob notes, which in those days was a good find. I telephoned the manager of the Bingo Hall, his comments were, thank you for being honest but you bought it and so the contents are yours. "Finders Keepers!"

Chapter 17

A CUSTOMER FOR LIFE

It was during the latter part of 1962 when I had a visit from a gentleman from London. His name was Thomas Neville; he was the managing director of a company in the City. The name of the company was G. Worrall and Son Limited and at this time they were operating at Salter's Court, in Bow Lane, E.C.4 They were a firm that employed highly skilled locksmiths, and had been established for over one hundred years. Safe opening and bank work were the backbone of their business; they also had an excellent counter trade. Our locks were suitable for new safes and for replacements in the event of a safe being opened and its existing lock being scrapped.

Mr. Neville had come to buy, and he was prepared to pay three times more than our current home market prices. This was quite appetising: we bit the cherry which turned out to be very fruitful. However, I would like to say a little about the Neville family because we became very good friends, and still remain so thirty years on. Mr. Neville was known to his associates and his entire workforce as 'Uncle Tom'. He was a well liked man with a great personality, he smoked like a trooper, and he was a social drinker who liked people's company immensely, he was a tough businessman, very tall and smart in appearance. Later on in my working life my involvement was with his son Ron, and eventually his grandson Mark, which resulted in a lifetime's work for me. Now during 1962 Uncle Tom asked me if I would be interested in the making of a special lock which they had patented. It was known as a Rack Bolt, or Pannier Bolt System and at this time a refurbishing programme was to take place at almost all of the branches of the bank and this lock had been accepted.

Although the job was to slow down during the next thirty years, which was in accordance with bank requirements, a service was essential throughout most of the country, as and when problems occurred, this made my working life complete and many of the locks will be in operation long after I have passed on.

My involvement in the Worrall Company had just begun. This is an introduction into the Neville Family, who are now and have been for many years directors of G. Worrall & Sons Ltd.

Chapter 18

A PHONE CALL FROM A BOOKIE

One day I received a telephone call from a Bookie in Birmingham. He had lost the keys to his safe and for obvious reasons the call out was most urgent. He was a very well known man in the business and operated a number of offices in the area.

I was to meet him at one of his offices exactly in accordance with what we had previously discussed on the telephone and it was important that I arrived on time, which I did. There was a Rolls Royce parked outside and I presumed that it belonged to the man I was to meet, so I went into the office to introduce myself. "I trust you have some identification", he said. "Yes", I replied and produced my business card bearing my name, and the name, address and telephone number of my company printed on it. "What other identification have you got?" He asked. "A driving licence, do you want to see it"? I said. "Yes please". So I went to fetch it from my van, and I gave it to him. He compared the signature, and name with that on the business card. "OK", he said "sorry about this but one can't be too careful. I've already had a load of trouble, that's why I didn't' give you all of the details on the telephone, you never known who's listening in". "Now I know who you are, we can go to my other office where the safe is kept". He said. So I followed him, he in his Rolls, and I trundled behind in my Mini van. At least we had two things in common, our vehicles were new and painted grey.

We arrived at the other office, he opened the door, took me into a back room and said, "there's the safe, how long will it take you to open it"? "I don't know", I replied, "I haven't looked at it yet". "Well get on with it", he said, "I've got to be at the races soon". I could see he was agitated so I opened it as quickly as I could. There were two Gladstone Bags, some books, and a lot of cash inside. He transferred the cash to another safe in the far corner of the room, locked it up, put his books under his arm, grabbed hold of the bags, and said, "come on let's go, I'm in a hurry, I've got to be at the races". "The job isn't finished yet", I remarked. "Well come back tomorrow and I'll pay you to finish it off", he replied. "Yes OK" I said. So he locked the door behind us, threw his bags and books into the Rolls and away he went.

The next morning I went back to complete the job, called at the main office and followed him again exactly as I had done on the previous day. After I had finished the work, he asked, "How much do I owe you"? "I'll send you a bill". "Not likely, he remarked, I'll pay you now". So I totted up his account, it worked out to seven pounds fifteen shillings, anyway after he'd paid me, I asked, "Have you got a good tip for today". "Yes", he said, as he put his right hand in his trousers pocked. "Here's ten quid and don't give it to another bookie or you'll lose it, and thanks for doing a good job".

I never had a tip like that again, and I didn't bet with it either.

Chapter 19

A VIEW FROM INSIDE

Our business friends in Birmingham were giving us plenty of regular work. I remember being asked to travel to Derbyshire on their behalf, where I was to present myself to the clerk of works at a large Police Station there. The building was under construction at the time, but was near completion and it was soon to be handed over to the local authorities.

The purpose of my visit was to carry out repair work on a cell door and grille gate locks. On my arrival I introduced myself to the site manager and asked to see the clerk of works. "He's not in yet, sit ya down he won't be long," he said. After a few minutes he came into the site office, we had a mug of tea and then we went to the cell-block to inspect and sort out the problems. We agreed upon the amount of work I had to do, so he gave me the keys and said, "when you've finished, will you hand the keys to the site manager". "Yes OK", I replied, and off he went. They were new locks, but some of them had been damaged, so the work that I had to do was going to take up most of the day. I had finished all the jobs except for one by late Friday afternoon. This last job was a repair to a grille gate lock, the gate led into a corridor, and to the male cells. I removed the lock from the gate and then I carried out the necessary repairs to it. After fitting it back, I tried the key in the lock from the outside and it worked OK. Then I made a stupid mistake. I put the key on the floor whilst I made a slight adjustment to the gate frame, and like a fool I forgot to pick it up before I went into the corridor and closed the gate. I was trapped, and it was my own fault. There was no exit. I called, "Is there anyone about, anyone there", but all was quiet, it seemed as if everyone had gone home. Surely the site manager must be about, I've got his keys, well I did have before I locked myself in here. It was getting late and I had to do something quickly. There were better places to spend a weekend than this, I thought. It was depressing, cold and miserable even during daytime. I felt like one of the bad guys in a cowboy film, locked in jail, looking at the key through the bars of the gate and couldn't reach it.

The key was about five feet away from me. Directly behind it was my box of tools. If I could lasso the box maybe I could bring in the key with it

and if a failed to get the key and could pull in the tools I would be able to get out anyway. I used my shirt and pullover, tied them together, put my arm between the bars of the gate and attempted to lasso the box. After a few times the pullover caught on a nail sticking up in the wood of the box and I was able to haul it in slowly. The key came with it. Finally I opened the gate, switched off the lights and got out as quickly as I could, and there was the site manager standing by his office. "I was just wondering how long you were going to be"? He said.

Chapter 20

IN THE NICK OF TIME

It was early one summer's morning, about 5-30, when I left home on a journey to Southampton.

During the late afternoon on the previous day we had received an urgent telephone call. This call was made direct from a shipping line. The instructions were for an engineer to report to the purser at his office aboard the `SS Pendenniss Castle`, where a safe was to be opened, the combination of the lock altered and new keys to be supplied. Also an early arrival was requested, this was of the utmost importance, because the ship was due to sail to South Africa on that day.

I was there early as arranged and after a brief talk with the purser, I understood from him that a key had been lost and also that the only existing key made available to me did not operate. However, after a few minutes routing inspection I soon came to the conclusion that the lock was not jammed, but that the key and lock was completely worn out. I tried to open it for some considerable time by using the original key but I was not successful so I abandoned the idea.

My next attempt was to try to pick the lock, but if I didn't succeed, then I would have to resort to the drill, which would create considerably more work for me. I had numerous sets of lock picks, some which were telescopic and were professionally made others, which were not. On this occasion I was successful, I had opened it within the hour, and then I removed the fire chamber and the lock from the safe.

My next job was to repair the lock, alter the combination and hand cut the required amount of keys. The engine room was available to me and any offer of help that I might require. I collected my tools and went down below where I was able to use a vice and carry out my work.

When I had finished there, I returned to the pursers' office. "How are you getting on with it"? The purser asked, "Alright I replied, I should be through in about half an hour". "That's fine", he said, "the captain would like to see you, and he has invited you to have lunch before you leave".

So I reassembled and completed the job as quickly as I could and then I tried all of the keys in the lock with the safe door in the open position and they all worked perfectly. Afterwards I closed the door, and locked it, I then picked up the old key by mistake put it in the lock and turned it. Obviously I realised what I had done, so I withdrew the key. However, when I tried the new keys again they wouldn't work, 'I had made a boo boo', and I was running out of time. I knew that if I didn't get the job done before the ship was due to sail, I would have to stay on board and come back with the river pilot. So I went down to the engine room, picked up a large lump hammer and a piece of solid timber. Then I went back to the safe placed the timber against the door face, gave it one almighty blow with the lump hammer, put the key in the lock and it worked.

My mistake was that when I altered the combination of the lock I had not realised that the original top lever had been filed to enable it to miss a cast ridge in the top plate. The keys I had cut were shorter on the top step than on the old key thus when the old key was turned in the lock it lifted the top lever to a high position trapping it behind the ridge in the cast top plate, so all that I had to do was to file the lever and the job was done. However, if the purser had done the same as I did, then the fault could have occurred at a later date. So once again I replaced the fire chamber and the job was complete.

I met the captain of the ship, my efforts were very much appreciated. I was given a super lunch and I left in the nick of time. My mistake was my secret and I learned by it.

Chapter 21

THE PRINCIPAL BOY

Harry and I came in from a job one evening, there was a message on my desk asking me to telephone a number. It was the number of a Birmingham theatre, and the manager answered my call. From his conversation I learned that his safe would not operate and that there was a need for him to lock cash away before leaving the premises that night. So I arranged to be there as soon as possible.

"OK Harry let's go, I've got a job to do in Brum", I said, "I'll drop you off at home on my way", "no" said Harry, "I'll cum with yer if yal buy me some fish 'n chips. If I go 'ome I'll only 'ave stew, cus that's all 'er knows ow ter mek" he said. So off we went to Birmingham. We hadn't gone far when Harry said, "do yer think we can stop fer a drink gaffer"? "No", I said, "you'll get plenty when you get there", "oh good", he said, "I'm a bit dry".

On our arrival we off-loaded the tools and I gave Harry my car keys and then I presented myself to the manager, There was the safe in front of me. I removed the fire chamber and the lock from the door. It was badly worn and in need of repair and a good service, which would probably take a couple of hours. During this time Harry had gone missing. I asked the manager if the bar was open, he said, "no it's not, but would you like a drink", "yes", I said, "I wouldn't mind a pint please". "OK what about your mate". "Oh he'll have one, he's only here for the beer", "well where is he?", he asked, "he has probably wandered off somewhere, or perhaps he has gone to the car". Anyway the manager brought two pints over one for Harry and one for me, but I was doing the work and I was thirsty and Harry wasn't back, so I drank the two!. It wasn't long before the manager fetched another couple of pints, which prompted me to go to the toilet, but when I got there I couldn't get in and I heard banging from inside, "who's in there" I shouted. "Who do ya think it is its me, Harry, arv bin locked in this bloody bog fer the past 10 minutes". It was late at night the panto had finished, but Harry was certainly caught with his pantos down. Anyway I went back to the managers office for some tools, opened the toilet door and out he came. "How long have you been in there", I asked again, trying to wind him up. "I bloody well told yer once", he said, "long enough, and if I'd 'ad a piece a wire wi me, I'd a bin out in a couple of shakes". "Well where have you been all this time "I asked, "I went ta the car" he said, "and ta try and get a pint, but the pub was shut". Anyway whilst I repaired the toilet door lock, Harry was knocking back as much beer as the manager would give him.

Chapter 22

THE ROYAL SHOW

During 1962, it was noticeable that the Withers Company was not doing so well under the new management, because we were continually being asked if we would take on more work on their behalf.

One of their most important local customers at this time was Mitchell & Butler's brewery of Cape Hill, Birmingham. The amount of new business that we were about to take on was considerable. We were to be involved not only in the distribution of safes to the Mitchell & Butler's Public Houses when required, but also the repair of safes as and when they became inoperative. Burglaries and attempted burglaries, obviously were included. The area we had to cover was mainly in Birmingham and Warwickshire and, as usual, working late hours was necessary because of the public house closure times. If however, a safe did not function, then it was essential that our service be prompt and reliable, in order that the licensee could secure the takings for the night. So we took on the work, it was another string to our bow and we did very well over the next three years with it.

Our first job in connection with the brewery was to renovate and alter the combination of ten safes; they were for the Royal Agricultural Show at Stoneleigh. We were asked to deliver and install them into the M&B public houses on that site; one safe of specific manufacture was to be installed in the Royal Pavilion where Her Majesty and Prince Philip could secure their valuables if necessary. After the closure of the show we would return to the site to collect the safes and put them into stock for the next year. Harry was in his element and put his name on top of the list for call out at any time; there was beer at the show, beer at the pubs, and beer at the brewery and all for nowt.

We continued to do well with the Mitchell's & Butler's work. It wasn't long after we had completed our first job at the Royal Agricultural Show when an enquiry from another firm of brewers came in. It was from the well-known firm of Arthur Guinness & Son. Their office was at Edgbaston in Birmingham, and if I remember correctly it was called Guinness House.

We were asked to carry out service work on locking securities there, to make regular check ups, and to be available on call should any emergencies arise. However, I decided to do the first job myself so that I could get familiar with the layout.

On my arrival there I presented my I.D. to a smartly dressed man wearing a dark uniform. His black tie stood out against an immaculate white shirt, his dark cap bore the shinning harp, the emblem of the Guinness business. "Come in", he said, and took me through to his office, where I was surprised to see three barrels perched on plinths side by side, there were also glasses ready for use. Obviously the content in the barrels was Draught Guinness. "So you'll be having a pint my lad", and he was quite quick to add that it was the best on tap around. "Yes please" I promptly replied, so we had a pint each, followed by another, and another until I was half cut. It was clear that I was no match for this fella, neither was I capable of doing the job in these circumstances, so I made my excuse, I've only come to introduce myself whilst I'm in the area", I said, "but I'll be here tomorrow at 9 o'clock", so I scarpered as quickly as I could and returned the next day as promised. However if I had sent one of my men he would never have got back and that could have lost us the business before we started.

Chapter 23

WORK FOR LIFE

Back at the factory it was all go. Withers' work was always wanted and was useful to us for cash flow, our export orders were progressing well at this time. Mr. Neville of Worralls Locksmiths, London, had the ability and personality for getting orders for lock work anywhere, and he was keeping us more than busy.

As I have said previously he was well known in the city. To me he was Uncle Tom, so I'll continue to call him that throughout my book.

We were at this time under a lot of pressure from him, especially for the rackbolt locks which were required by the bank. Working all hours, as usual, it was very difficult for us to meet the demand because there was a backlog, which we had to overcome before we could continue on an even keel. These locks were being made by hand in our factory; they were manufactured from best quality gunmetal castings.

The locking system operated two vertical rods, one to the ceiling, one into the floor, there was a horizontal latch bolt and locking bolt. The weight of the lock was approx. 10/11lbs, and the price to our customer was eighteen pounds per lock.

There were approximately two thousand branches of the bank. Most of these branches were to be refurbished, and we had got the order. The work was to be spread over many years, so to overcome the backlog was important.

At this time I had the foresight to see the possibilities of maintenance in the future years which became countrywide.

At the beginning we had our ups and downs regarding deliveries of locks but Uncle Tom and I were great friends and I liked his sense of humour. We despatched our locks to London by rail, carrier, and I myself made regular visits by road.

Whenever I made my visits I would be taken to the Whittington club in Bow Lane for meals and drinks for the day. In those days the laws of drinking and driving did not exist. I remember one incident very well. Uncle Tom and his wife, May, and the driver, Fred, were travelling from the North where they had been on bank business, and they were calling to collect locks from me on their way back to London. They had broken down at a small village just outside Buxton, and nothing could be done that night to get them away. So Tom telephoned to see if I could help. "Yes of course I can, where are you"? I asked, he gave me his whereabouts, and the name of a pub nearby, where I was to meet him. "It doesn't matter what time you arrive", he said, "We'll be in the pub". That was a foregone conclusion because they all enjoyed their drink. I packed a load of gear into my car, and off I went, it was winter time and snowing quite heavily. When I arrived there I found them in the pub as arranged, enjoying themselves. "Come on Fred", I said, "let's go and get the car hitched up and tow you back to Wolves".

We left Tom and May in the pub whilst we sorted the job out. I produced a piece of rope from my car and gave it to Fred, "you tie it on your end", I said, "and I'll do mine, and we'll be away in a couple of shakes". He laid out the rope on the snow and then he looked up at me in amazement and said, "What's this for"? "It's my tow rope", I answered. "It's not long enough to tie up my boots" said Fred. I remarked, "we're not tying boots up Fred, we're towing a motor, and that's all I've got, so the sooner we get off, the sooner we'll be home".

When we were ready to go I fetched Tom, and May out of the pub I had Tom as my passenger, and Fred was behind with May. He had a double hassle, a woman and a short rope. I motored as fast as I dare on the snowy roads, I wasn't worried about Fred he couldn't go anywhere, and he was a good driver. However we arrived back in Wolverhampton parked the cars on a hotel car park, and we were in time for some drinks. Fred was trembling like a leaf. May was terrified as she staggered from the car and said, "Get me a Brandy quick, I've nearly had a heart attack, and I never want another journey like that". Poor Fred was a bag of nerves, I don't think he ever forgave me, and I know he remembered his journey from Buxton to Wolves until the day he died.

Whilst the car was being repaired the next day we went to our factory to pick up goods, which Tom had to take back with him to London. The rest of the time we spent in the pub.

Chapter 24

THE LAST OF THE REB

Let's get back to the Union Reb, he was still working for the company, and his position as foreman had made him even more big headed than he was previously, if that was possible, his wages were well above the top rates, but like Oliver Twist, he always asked for more.

I didn't employ him because I liked him. He was disliked by everyone on the shop floor, his arrogance and aggravation was intolerable, but his work was good. Father would come and see me very often at this time and was interested to know how things were going.

He was working for a firm in Wolverhampton; they were a large private company known as James Gibbons Ltd., and were the oldest firm of locksmiths in the country (founded somewhere around 1650). They thought a lot of the old chap and he was happy there.

We were getting on together quite well now we weren't in business together, I could talk and listen to him, and his advice seemed much easier to take. Sometimes when we went for a drink together Father would say, "I don't like that Reb working for you, I know its not my business", he would say, "but that man will take all, don't trust him, get rid of him". "Well Dad I can't fire a chap who's doing a good job, and he is a valuable asset to the company". "Well you know your own know best, but don't say I haven't warned you". In spite of what the old fella said, the Reb was a bloody good workman and that's what I employed him for. It was now early June, 1962, when I was rushed into Hospital with an urgent appendicitis problem, and so the business had to be run by Maureen from the Office side, and the Union Reb continued to look after the shop floor, it was fortunate for me that I didn't take the old chap's advice on this occasion, because the Reb had temporarily become the backbone of the workshop and was a great help in producing Worralls locks for the Midland Bank. It was lucky for us that during the three weeks I was away, there was not much out work, and young George coped admirably. The gang who looked after the safe removals and installations were mainly casual workers, and carried on as usual.

After my three weeks illness I returned to the factory, not to work, but to potter about and to see that things were running OK. I was more than pleased to find out that everyone had done a great job whilst I had been away, and my appreciation was shown by giving my congratulations and a bonus. Also a special bonus for the Reb who had looked after the production side so well in my absence.

During the day I telephoned Uncle Tom, to have a chat, and tell him that I was back in circulation again. "Oh great", said Tom, "How are you feeling"? "Not so bad", I said, "I've got to take things easy for a while", and so we carried on with our conversation, then he dropped a bombshell. "Are you well enough to take a shock"? He said, "Why what's the problem Tom", I asked. "Well" he said, "You've got a traitor in the camp", I was taken back for a minute, "why w-w-what do you mean Tom"? "Well last week your foreman rang me to tell me that he had despatched my locks, he also said that he had a proposition to put to me". "Oh", I said, "what sort of proposition Tom"? "Well I'll tell you but in strict confidence", "OK I'm listening". "He knows where there is a workshop for rent somewhere in Wednesfield" "Oh yes" I replied. "He's after trying to take over your work, also he has a locksmith who is ready to work with him". "Well Tom", I said, "this is a bolt out of the blue. What did you tell him"? "I told him that his gaffer was not dead, but was ill, and that surely he must understand that we are in an honourable profession and I do not make deals with the likes of him or his type of person". I thanked Uncle Tom for telling me, and added that I must think about it, and deal with the problem in my own way. I felt very hurt about this, because the old fella was right again, I was employing a snake, and I had to get rid of him the sooner, the better, but how was I going to do it?

In those days the trade unions were very strong, also Uncle Tom also had told me in confidence, and I had no evidence of the Rebs intentions, but I was aware of the possible repercussions, which could arise within the union of which the Reb was a staunch supporter.

I telephoned the old chap to tell him what had happened. "I told you, didn't I"? He said, "get rid of him as quick as you can". So I got thinking, I had to be as crafty as a cartload of monkeys. I knew that the Rebel was going on holiday to Devon for the first two weeks in July. The other men apart from George were having their holidays directly following the Rebs return. During the two weeks whilst the Reb was on holiday I made my plans to get rid of him. I had all the work in progress finished off, and despatched, including exports to South Africa. New orders were coming in, and we had plenty of existing orders on our books. These were kept under lock and key in my private filing cabinet, this gave me an advantage on what I was about to do.

The day that the Reb was due to return home was on a Saturday, so I went to his house early that morning, I put a note through his letter box which read as follows: -

"I will be in work tomorrow, Sunday, would like to see you".
Signed W.B.S.

He took the bait, and came on the Sunday morning. "Wot da ya wont", he said, "I av'nt finished me 'ollidays till Mundy". Little did he know that he was going to have some more holidays in the very near future. He looked around the workshop, and commented, "somebody's bin working 'ard, all the works gone". "Well", I said, "we have to get the money in to pay for the holidays". "OK, wot da yer wont me fer." "I've got to put you on three days a week until we get more work in". "Why me", he asked, "wot about the others". "They will be on three days when their holidays are over". I replied. "Oh well I suppose I'll after put up with it, but I 'ope yer know arv got two kids ter keep". I was relieved because I had got over my first hurdle, although I was hoping that he would have given his notice there and then, or have asked for his cards and gone. However, he stuck it out for three weeks, and I was on edge all of that time, until one day when he made his first move, and said, "I'm leaving gaffer, arv got me another job, I ain't putting up wi this any longer, so ya con tek a wick's notice". Another employee came to the office to see me during the afternoon on the same day, and said he was going to leave. I had noticed that the Reb was quite thick with him, and had been for some time, which was unusual, because the Reb was a bit of a loner. However, looking back was it possible that he was the locksmith who was going to join forces with Reb and try to take over my work? I think so, because they both left together, and went to work at the same firm.

That was the last I heard of them.

Chapter 25

TAL-Y-BONT, TAL-Y-BONT

It was quite a few weeks before I managed to find a replacement locksmith, because the type of workman I was looking for had to be very highly skilled, and they were very few and far between. I decided not to replace my foreman, which meant that I would look after the shop floor myself. Our female workers were good, and providing that I set the tools in the hand presses, and the power press, there would be few problems.

My life was hectic and I was always on the go, seven days a week, and as I have said before, working all hours. I was buying one hundred cigarettes a day at this time, and I was smoking sixty of them. I left fags burning on the office desk, also on machines and benches in the factory. George and I worked together on emergency calls, of which there were plenty; also this type of work was on the increase. I had to fit all of my work in accordingly, and at times the clock had no meaning to me.

I remember one incident in particular involving George; it was when a call came through from George Masons Head Office in Birmingham. A safe had jammed at one of their branches, the branch was at Tal-y-bont, I wrote the details on my work pad, gave a copy to George, he stuffed it in his pocket and away he went. Some three hours later after George had gone, we had a frantic phone call from him, "I'm at the wrong bloody place", he bellowed, "put the gaffer on Maureen". "Hello George, what's the matter"? "I'm in Tal-y-bont, and there ain't no bloody Masons eer". "There's gotta be", I said. "There ain't," he said again, "I'm eer and I bloody know". "Just a minute George, I'll check". So I checked my copy, O hell, I hadn't put on which Tal-y-bont he had to go to, and there were three. It was my fault and he wasn't very happy when I gave him the correct location. "I'm bloody miles away from there", he said, "and I ain't gonna get 'ome tonight". "Well George", I replied, "If you had looked at your map index properly you would have queried it with me". "Oh no, you gaffers are all the same", he said, "you ain't gonna blame me for your mistakes, it's your bloody fault, and I've got no time to waste, so cheerio", and he carried on to the right Tal-y-bont.

Chapter 26

"EE COR BE"

Shortly after the Tal-y-bont episode I sent George to Bristol, it was on a Wednesday and he liked his Wednesday evening off, because he always took his girlfriend Jose to the pictures. On this occasion he rang through just before lunch to say that there was a lot more work to be done than was first thought, and that if Jose rang would I let her know that he would be very late getting back. That meant no pictures for Jose and she wasn't going to be very happy. Now Jose came from Walsall and had a broad local dialect, she worked for a very old firm of saddlers there, they made saddles for royalty all over the world, and for film stars such as John Wayne etc. anyway on her way home from work she rang me. "Ello Bob, con-ar spake ter George", "Sorry he's not in, Jose". "Well wee'r is e? I 'ope ya ain't sent 'im on one of them escapades again". "Yes he's in Bristol" I replied. "Ee cor be, e's tekin me ter the flicks tunite". "He won't be back until very late", I said. "Ee's garra be back it ee's tekin me out". "Well Jose" I replied "you'll have to do without him for tonight". "Wait till I see im I'll ge im Bristol, ter rar", and off she went. Whatever she gave him when he came back from Bristol must have been very good because they soon got married.

Chapter 27

ONE STEP AHEAD

Over the years we had made locks for a firm of safe-makers known as the Birmingham Safe Company. They were originally one of Father's early customers; they operated their business from Heneage Street in Birmingham until eventually they were taken over by a supposedly wealthy business man known as E.J.C. His family were in the business of manufacturing push bikes, and their product was named Hercules, they were one of the leading companies in that field at the time.

After the take-over the Birmingham Safe Co. moved to a large site which was known as the Pensnett Trading Estate, near Dudley. They had an elaborate set up there, a spacious shop floor and luxurious offices. It was far in excess of the premises previously occupied by the company. They continued there for sometime, and we were still making locks for their safes as we had done in the past, but their debt with us was accumulating. The excuse for not paying was cash-flow. However we were assured that our money was safe, and the manager Mr. Palk, told me that cash was to be injected into the business by the Uncle of E.J.C. who was a director of the Hercules Cycle Company and lived on the Channel Isles.

However, there was no sign of any settlement of our account, and we were still supplying our locks. I was quite friendly with Mr. Palk the manager. He would regularly telephone me begging for locks, which I had in stock ready for their safes. There was no way in which they could despatch these safes without our locks because they were jigged up to our design. I remember one day when Palk telephoned. "One hundred and fifty safes waiting to go out, and no locks" he cried. "What about it"? "What about our money", I said. "You haven't paid for months". "Well" he replied, "bring over all the locks you've got and we'll give you a cheque". "OK I'll come over this afternoon at about three o'clock", it sounded convincing, so I put the locks in the boot of my car and locked it up and as arranged. I arrived there at three o'clock. I went to Mr. Palk's office where he sat behind a posh desk and he shook hands with me on my arrival there. "Have you brought the locks", he said. "Yes I answered, and here is my account to date". He pressed the bell on his desk and in came the foreman. "Ah

Jack", said Palk, just get the lads to collect the locks from Mr. Sidbotham's car. I interrupted, "Excuse me Mr. Palk, my cheque first, please", then I continued, "the locks are in the boot of my car, and it's locked". "Well let's be reasonable", said Palk," we may as well offload the locks whilst your cheque is being made out". "Firstly no Mr. Palk, secondly, I am a very reasonable man, that's why I have given you so much credit in the past. Anyway if you haven't got a cheque for the full settlement of the account I would like to see your managing director". "Sit there a minute", he said, and he went out of his office. A few minutes later he returned, "unfortunately E.J.C. has gone", he said, "and he won't be back to day". "Oh then he must have gone without his car, because I've parked my car in front of his, so if he's in he can't get out". Palk made another exit from his office, and returned again after another few minutes, to say, "Sorry I thought E.J.C. was out, it seemed that he had gone missing in the factory, he's coming to see you". So after about ten minutes in walked the dapper little director E.J.C. "Hello Mr. Sidbotham, sorry I didn't know you were here, how are you"? and he went to shake hands with me. "I'm sorry Mr. C.," I said, "but my business is not of a pleasant nature, I have come to collect my money, and if there is no money then there are no locks", "right", said E.J.C. to Palk, "have you got the WBS account there"? "Yes sir" replied Palk. "Then pay the man", he said, "and good day to you Mr. Sidbotham", and he walked out.

I received the cheque for the full amount from Palk, and gave him the locks. I left feeling quite satisfied, I expressed the cheque the following morning. The cheque was no problem, but I was very lucky because the official receiver was appointed the following week.

Chapter 28

MY LUCKY ESCAPE

After a few weeks I had a visit from Mr. Palk. He had been dismissed from the Birmingham Safe Company, and was out of work. When he came into my office, he insisted that I called him by his first name, Archie. This was a little different than when I last visited his office. Now Archie was a pudding of a man, bombastic, a drinker with a pot belly, but a likeable chap. He was clearly depressed, his marriage had gone wrong, and he needed some work so that he could earn a few bob: "anything will do just to keep my mind occupied," he said. After all he was manager material, and used to dealing with workmen. "I'll give you a job", I said, "you can take charge of the safe removal and installations side of the business". He accepted straight away, even before I had discussed any payment, and we became good friends.

It was fortunate for me that I gave Archie some work because I became very involved with the West Midlands Gas Board in Birmingham, when many of their branches were closing down, and, apart from safe removals, they were selling off safes, desks, filing cabinets, and other types of office furniture. I was asked to make offers as and when these kind of goods became available, and of course I had the moving power, so I purchased at nominal prices.

I brought the furniture back to the factory, and this meant more work for Archie, he was to renovate and paint it ready for despatch to the local auction rooms in Wolverhampton where it was sold for a handsome profit.

I could rely on Archie, he was a good safety man, and he would check the setting up of our equipment that was vital, especially when moving heavy safes up and down stairs. But there was a day when Archie was ill, and I had to take the gang to Birmingham. I was to move safes from the upper floors to the ground floor at Saltley Gas Works. The stairs were made of stone, and there were three flights, so we set up our equipment, and moved the smaller safes first. The weight of one of the larger safes was approximately one and a half tons, it was a double door about six foot six inches high by five foot wide, an old safe, but of good quality, and was made by one of the leading makers at the time. The staircase was made of

concrete and was very wide, we were moving the safe downstairs on its
back from the first floor, I was on the stair just below the slowly
approaching safe, which was controlled by a ratchet winch, when suddenly
I heard a crack, the safe had come adrift. In a split second I jumped onto
it, scrambled up and gripping onto the top ledge, it was like travelling on an
express train, I had the presence of mind to stop my feet from overhanging
its base when eventually it crashed to the tile floor, where I jumped off.
Finally it came to rest by damaging the brickwork in the facing corridor
wall. Another stupid mistake, I should have checked the safety chain had
not been secured properly, also the ratchet on the winch had failed. It was
a close shave and a lucky escape for me.

Chapter 29

WORKING WITH THE LAW

East Anglia was quite a journey from our factory in Wednesfield, and during the Sixties it was not one of the easiest of places to get to. On this occasion my call out was to Great Yarmouth. There had been a robbery at an engineering company there. The managing director of the firm claimed that a substantial amount of cash had gone missing from the wages safe, and he asked if I could visit that same day. I was either to fit a new lock, or alter the combination of the existing lock and provide new keys. I accepted the job, and within a few minutes I was on my way. It was about five o'clock in the afternoon when I arrived there. After producing my I.D. I was taken to one of the director's offices, where I was asked if I had any objections to a detective constable being called in to ask questions and make observations. I already knew that the Police had been there most of the afternoon, and although this was an unusual request I was quite happy to go along with it. So I waited for the constable to arrive. During this time I was taken to the manager's office where the safe was housed.

I acquired considerable information from the manager who wanted to talk about the incident; he seemed to be nervous, and understandably very worried. Firstly, he told me that the money stolen from the safe was a large amount, and that most of it was for the wages, also that the cash had gone missing between eleven a.m., and twelve noon whilst he had been out. He said, that he had locked the office door before leaving, and when he returned he found that the office door was open and that the lock had been forced, but the safe was locked. "So what did you do then"? I asked. He said, that he reported to his director, who then came down to the office, and they opened the safe but alas the wages had gone. Anyway our conversation came abruptly to an end when the constable arrived. I was introduced to him, and then I was able to start my work. The safe was open, but it was not open at twelve noon when the manager returned to his office and reported the break in. Anyhow I proceeded to remove the fire chamber from the back of the safe door, when the constable asked me a question. "Is there anyway that you can tell if a lock has been picked"? "It is possible", I replied, and I continued, "in some cases where picking tools have been used you may find, by looking carefully, on both faces of the levers, irregular marks which could indicate the use of such tools. However if you are thinking that this safe may have been picked, then forget it". "Oh" he said sharply "and why"? I replied, "because the door was locked after the money had gone, and the manager didn't know until he opened the safe with his key. At least that's what he told me, but if the safe had been picked and the contents removed, then its hardly conceivable that a robber would put the bag of SWAG on the floor, and try to lock the safe by the same method". I continued to examine the lock and there was no evidence to suggest that the safe had been opened with any implement other than its key. At this point the constable was about to leave. He had been most interested and thanked me for his first lesson in locksmithing. I was asked if I would make a visit to the Police Station before leaving for home. So after finishing the job, as arranged, I called at the station where I was asked for my opinion as to what had happened, and for a report on the technical observations I had made. I gave a statement, and a verbal opinion in my capacity as a master locksmith that there may have been a duplicate key cut to pattern, or that keys could have been lying around long enough for someone to take the opportunity of making an impression. Was it an inside job? I never knew, but I had a good meal in the Police canteen before continuing on my journey home where I arrived at about four o'clock in the morning.

Chapter 30

JUST DESERTS

It was late one afternoon during 1962 when Archie had mustered up the gang and we went to move a safe in Willenhall. The safe was a two door, six feet six inches high and five feet wide, and its weight was between one and one and a half tons. In fact it was very similar to the one that chased me downstairs at the West Midlands Gas Board, but this time there were no stairs involved. We were to move it from the customer's old premises because demolition was to take place during that week.

The new premises were under construction, there were doorframes fitted, but no doors, and neither was there a roof on the premises, so the place was open to all and sundry.

On the floor in the corner there was a rectangle marked in chalk where the safe was to stand. We were provided with a tarpaulin to cover it up. After completing the job we all left the premises and went to the pub for drinks, and afterwards at about ten thirty we were off to the fish and chip shop, and then home. I went to work the next morning, and at about nine thirty the office doorbell rang. I was in the workshop when Maureen called me. "Bob, there's someone to see you". "OK", I replied, "coming". There were two men at the door wearing trench macs, obviously they were detectives, or looked like detectives. "Good morning", I said, "what can I do for you"? "We are from Willenhall Police Station, can we come in"? "Yes OK", I replied, and so the questions began. "Can you tell me where you were last night"? said one of them "Yes I can", we were moving a safe in Willenhall, why what's the problem"? "We will ask the questions if you don't mind", said the other. "Where did you go after moving the safe"? "To the pub", I replied. "And what time was that"? He asked. "About nine o'clock", I replied. "Which pub"? "The Windsor Castle". "Yes OK, and what time did you leave"? "About ten thirty". "And then"? "We all went home". "No you didn't", he snapped, "you went for fish and chips". "Oh yes", I had forgotten about that. "Can you vouch for all your men"? They asked. "Yes", I replied, "until they went home". They paused for a while, so I asked my question again. "Well now officer, can you tell me what this is all about"? "Yes I can", said one of them. "Firstly we know the names of your crew, secondly, the safe that you moved last night, there was an attempted break in". "Oh hell, I'll help all I can, but what can I do"? I said. "You can come down to the station with the men that took part in the removal of the safe, this should eliminate them from our enquiries". "Yes I understand, two of them are here in the workshop, we'll come right away". So we went down to the nick to have our dabs taken. I didn't ask any more questions until we were cleared, and then I was told that thieves had broken into the old factory during that evening, they had stolen the oxy-acetylene cutting gear, taken it to the new place, and cut a great hole in the safe door, but failed to open it. What a right set of charlies they were to have thought that there would be cash in a safe, in a building that was open to everyone. They must have worked bloody hard for nothing, and they got their just deserts. Later on we were called into open the safe and repair it when we found a grand total of seven shilling and fourpence halfpenny in one of the draws, and that would have been their haul.

Chapter 31

WHITE'S POP

Another attempted break in was at White's Lemonade factory in Nottingham. There were three safes there, two which we had to check over, and one that the intruders had tried to blow; which had resulted in its keys being inoperative. So George and I humped our kit in the car, and off we went. When we arrived there, police photographers, and dabs men were invading the place. "Have you finished working on this safe"? I asked. "Yes", he said, "you can have it, it's all yours", and they carried on dusting round the office.

Now George didn't like working with a bunch of cops bustling about, also he was a bit of a devil may care sort of bloke. "I ain't working with this crowd hovering around me", he said. "OK then", I replied, "let's go and fetch our kit, they've nearly finished". "No" he said, "you go Gaffer, and I'll try to get rid of 'em, they only want to watch us". Well somebody had to go and get the tools, and I guessed he was up to something. I came back after about five minutes to find that all the cops had gone into another room to work. George was looking into the keyhole in the safe to see if he could find any obstruction. "What have you done to them George, they've all cleared off"? Just then one of the cops poked his head round the door and said, "your mate's bloody daft, the silly bugger's gone and struck a match in front of the keyhole", and he continued, "there might be some jelly or a detonator in there". "Well" I said to George, "what a bloody soft trick to do, you've got your torch, why don't you use it"? "Ah he replied, but I told you I'd get rid of em, didn't I"? "Yes" I said, "you could have blown us all to kingdom come". George had made a mistake, and I knew he would learn by it, so I thought it best not to pursue the matter further. We continued with our work, and decided that we would have to use the drill to open the safe. Now if the Police had been right, we could have been in considerable danger, and White's pop could have gone up with a bang, and all of us with it. We had to be very careful, because the position of the hole that we were going to make in the safe door was to be to the right of the keyhole, which meant that we were drilling with the left hand, and working on the right of the machine keeping as far away from the keyhole as possible. Drilling had

to be done slowly, and with care, especially when we entered the lock mechanism, because we had no knowledge of what, if anything had been put into the lock. Anyway all went well.

After opening the safe we removed the fire chamber. There was no detonator, but in the bottom of the chamber we found a small amount of explosive material which may have fallen through a slot in the wall of the lock cap. George played safe this time; he took it outside fired it with a match when it fuff't up and fizzled out.

Chapter 32

A JOB TO REMEMBER

It was during 1963 when I was called out to the Dunlop Rubber Company at Fort Dunlop, Erdington, Birmingham. There was a large amount of cash involved and it had to be locked up overnight in the strong room. The door mechanism had jammed in the unlocked position. I had worked at the Dunlop factory on many of their securities in the past and I was familiar with the job we had in hand.

It was a very old door and frame manufactured by the Whitefield Safe Company, made during the reign of Queen Victoria and dating around 1875-1900. At the time of manufacture it would have been quite highly rated. Although I had serviced it many times in the past it was now almost worn out and living on borrowed time. So we had to get it working.

The strong room itself was well constructed, and the door and frame were of matching quality. It was a main security location for wages in Fort Dunlop at that time, and we were told that there was about a quarter of a million pounds deposited there. The locking mechanism was worked from an outer spin wheel, and the door was secured by an eight-lever lock. The inner part of the door mechanism worked from the spinwheel to a set of cog wheels and a rack, and eventually to the main bolts which were worm threaded and mated with the worm threads in the door frame bolt holes. When the wheel was turned to its fullest extent the key would operate, but in this particular case the door had dropped on its hinges thus resulting in the main travel bolts becoming jammed against the worm threads in the door frame which prevented them from travelling their full distance. However after a few hours stripping down and making adjustments Harry and I decided that we would give it a try, so we reassembled the mechanism back on the door. When we closed the door and turned the spin wheel the key would not quite operate the lock bolt. The door was fouling against the door frame so we proceeded to grind the leading edges of the door. It was almost there when we closed the door again, but not quite. So I put my torch on and went inside, I asked Harry to close the door slowly and then carefully turn the spin wheel. "OK Harry", I shouted, not thinking that he could hardly hear me, "I think I've found the problem". I thumped on the fire chamber, "open the door, open the door", but he didn't, then I realised that he had got problems too. Although the door had not been locked with the key Harry had turned the bolts into the locked position and obviously they were jammed. There was a lot of banging going on outside I must have been in there all of fifteen minutes, it seemed like hours, fortunately I didn't suffer from claustrophobia, but, I sweated buckets, whether it was from heat or panic I don't know. Anyway eventually the door was opened Harry had broken off the spin wheel. With the aid of a Stilson wrench and a big lump hammer he was able to turn the spindle, thus withdrawing the worm thread bolts from the doorframe, after which we continued with the job, much to the satisfaction of our customer. It was certainly a job to remember.

Chapter 33

DECEPTION

How easy it is to become a sucker for a hard luck story, especially when some poor down and out old dear calls on you.

My office bell rang, I opened the door, and there she stood in the pouring rain. A little hawky type woman, almost in tears, I thought, or was it the rain running down her face. She must have been in her late seventies or even older, a poor soul, looking drab and almost in tatters. I always remember how sorry I felt for her, "come in my dear you must be soaked" she came into the office, and I gave her a chair, "what's the matter"? I said, "Can I help you"? "Yes I hope you can", she replied, and continued to say, "I have just had a bereavement in the family, and I have got a little safe, but I can't find the key, and I haven't got any money other than a few shillings which are in the safe".

Well I was a soft touch anyway, and said to her, "don't worry love, I'll come and open it in the morning, and if you wait a few minutes I'll get you a cup of tea, and then I'll take you back home". So I took her to Willenhall. The house was a pokey dark little place, and about as drab as she was. A young chap opened the door, she went inside, and I followed, I had a quick look at the safe, and said, "I'll come over in the morning to open it and I'll cut you a couple of keys OK"? I returned the following morning as I had promised, I was there for about two - two and half hours. When I was ready to go, she half-heartedly offered me two shillings and sixpence for a drink. Which she had taken from the safe. "No thank you my dear". I said, "Have this one on me". She didn't offer again, but she did thank me as I came away. A couple of days later I mentioned it to Harry, because he lived in Willenhall, and he knew a lot of people there. When I told him the woman's name he said, "you've bin bloody 'ad, she's a money lender, and er's got more money than you'll ever 'ave". Then I told him about her bereavement. "Bereavement to buggered", he said, "You've bin conned". A few years later I learned the old money lender had been brutally murdered.

Chapter 34

A BUDDING CRIMINAL

We had another trainee at this time, his name was Nick and he was progressing with the job very well. A hard worker and way above the average considering that he hadn't any previous experience.

It was after he had been working for us for about twelve months that I noticed a change in his attitude towards the job. This gave me a gut feeling, and without reason I just didn't trust him anymore. He was always very interested in the making and filing of keys, in fact more so than the working of locks, which was quite unusual, and it puzzled me because most of our youngsters had shown more interest in the lock mechanism than the key. Firstly because the keys were cut for the workmen, by myself, secondly the gauging and rounding of the key steps was very monotonous and accurate work, especially in hand made locks, which were not on a production line. The lad was quick to learn and asked a lot of questions, some of which were very suspicious, such as how to make skeleton keys.

As time went on I became more uneasy, and aware of the fact that there was some pilfering going on in the factory. Sometimes, I would notice that cigarettes had gone missing, and also small quantities of brass had disappeared. However there was nothing I could do at the time other than to lock the fags in the safe and keep quiet.

Although I had no reason to be suspicious of the lad, I had to find out and I did have that gut feeling. I kept a keen eye for two to three weeks and small quantities of brass still continued to go. Anyway something happened which was quite irrelevant to the situation that I was trying to solve. One afternoon, the lad came back from his lunch hour. "I'm sorry I'm late", he said, "My mum's had to go to hospital". "Has she", I asked, "is it serious"? "I don't know until I go to see her". "Well, when can you go"? "This afternoon", he said. "OK then, have the afternoon off". "Oh thanks", and he went. The next morning he was late coming into work, but I thought well, he's probably got some housework to do. When he did arrive, I asked, "How was your mother yesterday"? "She wasn't very well, but she was a bit better when I saw her last night". He said. "Is there visiting this afternoon"? "Yes"? "Very well, you had better do the same as yesterday, and go to see her".

I didn't give him any more time off after that, but I continued to ask about his mother. Anyway a couple of days later, at about two o'clock in the afternoon I had a phone call from a Dr Bennett at New Cross Hospital at Wednesfield, "Can I speak to Mr. Nick Jones"? "No he's not here at the moment", I said. Then the bloody penny dropped, it was a bogus call, this was no doctor and I knew it, but I continued talking, "can I give him a message"? "Yes please", came the reply. "Will you tell Mr. Nick Jones that his mother is not well, and can he visit her this afternoon". "Yes", I said, "leave it with me, he *will* be there". It was a bogus call, but I had to be sure and check it out. So I telephoned the Hospital, there was no Dr Bennett there. In the meantime Nick Jones had come into the workshop. Whilst I was on the telephone, I was watching the lad through my office window when he looked up expectantly. I went into the workshop and gave him the message and said, "come on, get your jacket, and I'll take you in the car". "No it's OK I'll go on me bike". "No" I insisted "I am taking you, so get in, and I think you're a bloody liar". Clearly I shook him; I just hoped that I was right. It was about a mile to the hospital, and he stuck me out all the way. I began to worry especially when I parked my car on the hospital car park, and he got out. He walked toward the long corridor, and I went with him. This ones a brazen begger if ever I've seen one, or I'm totally wrong, in which case I've dropped a right clanger. But as we got to the corridor he suddenly said, "OK I lied". What a relief that was to me, I had bluffed the bluffer! I got him back to the factory gave him his cards and told him to get out. I never found out who nicked the fags or the brass, but we didn't miss anything else after he had gone.

Also it wasn't the last I heard of him, because later I heard that he was in trouble with the law for breaking into sheds with the use of skeleton keys that were probably made under my factory roof. No wonder that he was more interested in keys than locks.

Chapter 35

LOOKING FOR A COTTAGE IN THE COUNTRY

Things were going very well, and we were getting over the backlog of Uncle Tom's rackbolt locks for the bank. Financially we were also on the upgrade, and work continued on an even keel much more than it had in the past, which meant that I could snatch a bit of time off during weekends. There was something missing in my life and suddenly I realised what it was.

I was living with my in-laws. I had no hobbies because I hadn't time for any, my mind was full to the brim with business. I had shut everything out of my life, even my wife and daughter. At this time I saw myself as being a total failure to everyone including myself, all for the sake of the bloody business. I had to get my priorities right. I knew that my family was more important to me than anything else in the world, but I brushed them aside. Moving from my in-laws had now become my first consideration, and we were not short of a bob or two.

Both Ann and I wanted passionately to live in a country cottage, preferably in a remote area and amongst the farming community, but to find what we wanted was going to be difficult. A mortgage was out of the question, especially on old property. But if we could buy a derelict cottage, or an old mill, or anything that we could convert, or renovate, then that would be just great. For us to do this any other way would have been impossible, because our finance was limited.

So we began our search, which had to be somewhere within a radius of about twenty miles from Wolverhampton, preferably in Shropshire or Staffordshire.

We found many cottages for sale, but there were always snags, such as no water, no electricity, or both. Some of the places we found were condemned, others were too expensive. Although we were going to draw the money from the business to buy our property the expenditure on materials afterwards would be very costly and there would be no mortgage available. We continued to search throughout the summer of 1963 when we came across an old mill which had been driven by a water wheel. There was a miller's cottage there, but again it was condemned. We tried to lift the order and failed. There were plenty of tumbledown properties and many of them were beyond repair. However there was one particular cottage advertised in the local newspaper, which appeared to be quite interesting. So we made our enquiries to the estate agent in Wolverhampton.

It was a country cottage for restoration, at a small place known as Giffords Cross and it belonged to a private estate nearby. It was a crofter's cottage, thatched, about four hundred years old, Elizabethan, and constructed from small red brick. Its thatch had caved in, and it was somewhat of a tumbledown wreck. The terms of the agreement were, that the successful applicant (tenant) was to restore the property at his or her expense and maintain it. Also the property was leasehold. In return the tenant was to pay the fixed sum of ten shillings per week rental throughout the lease. Each applicant was to submit plans together with a finished drawing of what the property was going to look like on completion.

After thinking about it for a couple of weeks we decided to give it a go. So we submitted the plans and our drawing, which I thought was very good, especially coming from a locksmith.

During the next twelve months we telephoned the agent on numerous occasions and got nowhere, and the old cottage eventually became a heap of rubble over the years, but we kept on looking for the good life, and found it twelve months later.

Chapter 9 - Kitchen Gadgets

Chapter 14 - A Works Trip To Ascot

Chapter 19 - A View From Inside

Chapter 28 - My Lucky Escape

Chapter 31 - White's Pop

Chapter 43 - Moving Out, Moving In

Chapter 36

A BOOZY DO

Uncle Tom came down from London for a weekend to a birthday do which was for one of his old associates. His name was Hughie Price, and he was known to his friends as Bunker, he was a factor dealing in locks and keys. Now Bunker had a lot of boozers around him and they were invited to the do.

We met at his house at nine o'clock one Saturday morning, there were eighteen of us in all and we were asked into his house for drinks. None of us knew where we were going on that day, but we did know that we would enjoy ourselves. After we had been drinking at Bunker's place for about an hour, up rolls a coach and it was for us. So we piled in and we were on our way to somewhere. We stopped at Llangollen and other places to use toilets it was a good excuse for a pint. Eventually we guessed that we were going to Llandudno we were right, our coach pulled up at the Grand Hotel car park. We all jumped off and went to the dining room for a good nosh-up which was already booked for us. During the late afternoon we left the Hotel taking the scenic route for home, which took us over the mountain roads where we came to a little pub tucked away in the hills. There was nobody in the pub when we arrived. The place didn't look very prosperous, but the licensee was pleased to see us, and it was obvious that she could do with our custom.

The night was young and we were going to spend a lot of money there. She lit a big fire, and we were as comfortable as bugs in rugs. Although Uncle Tom was a locksmith, he was also a good entertainer, so we soon got going, Tom's wife May was on the piano, and Tom started the ball rolling. We were soon getting well oiled, and our singing was loud. I remember it became very hot in the pub, so we opened all the doors and after a little while one of our fellas went outside for a breather, he hurried back in, and said, "come outside here chaps and listen". There were folks coming up the hillside from the Valley the echo was fantastic as they sang there way up to the pub. "Why don't we join forces?" one of them said. It was a wonderful night, we all sang together until one o'clock in the morning; some of us were inside the pub, and a lot were outside in the open air.

It was a great day to remember.

THE TURNING OF THE KEY

We take most things for granted
It's all so plain to see.
The one that's ever in my mind
Is the turning of the key?
Taken from our pocket, a handbag, or wherever,
We place it in the lock. To open up the door,
Suddenly it's jammed not working anymore
Twenty years of service
That's what it's all about
Why do we not accept the fact that the lock is now worn out?
Frustration and panic, hits us naturally,
Because we take too much for granted in the turning of the key.

PART II

A LOCKSMITH
TAKES TO THE COUNTRY

Chapter 37

A LOCKSMITH TAKES TO THE COUNTRY

It was during the spring of 1965 when Ann and I decided that we had to make up our minds about moving into the country, so we searched the Staffordshire and Shropshire countryside for old barns, water mills, and more tumbledown cottages. All were either crumbling with decay or not for sale.

"FOREST EDGE"

A Cottage at the Forest edge,
    Three hundred years, or more, hath stood.
Now crumbling with decay,
    As night draw's near,
It's views fading for yet another day,
    With timbers dislodged its walls abow.
I wonder,
    "What future", hath this cottage I know,
I listened; I waited, as time passed by
    The moon was full,
The cottage, a silhouette against the night sky,
    The valley below was embraced within a shrouded mist.
Owl's hoot, a curlew's call,
    Cricket's chirp within a sandstone wall.
For many years, I've walked this way
    To see the cottage, the tumbledown cottage at the Forest edge,
in it's state of disrepair,
    and increased dilapidation.
For now I'm told, it's just been sold.
    With conditions for restoration.

It was on a glorious Sunday afternoon during April 1965 when fortunately for us we took a wrong turning and got ourselves lost amongst the narrow lanes around the Norbury, Newport area, and we ended up between Aqualate Mere and Norbury Junction in Staffordshire.

The Long Boats on the Shropshire Union Canal were beginning to perk up for the summer season. Put-put-put went the diesels as they crept forward into the boundaries of Shropshire and Staffordshire. Little did I know that my country life was to be so involved with the sound of the diesel engine? It was great seeing people enjoying themselves, boating trips later became very popular, and of course, holidays on the canals were here to stay.

I began to wonder why I had worked all the hours that God sends and let the rest of the world go by.

This was certainly a great place to live, and so we trundled off down the country lanes in search for an old cottage that we hoped would become our home. We hadn't gone far before we found one, and although it was a wreck a chat with the owner wouldn't be amiss. 'Sorry', said the farmer, "I'd like to sell it to you, but I can't it's a family heirloom, and it's condemned" he said as he pointed to a huge structural crack running from chimney to ground. "Sorry can't sell it" he chuntered on as we left.

A few seconds later a fella came along on his bike. He was a local chap out bird watching, whom I later became to know as Stan. "Excuse me", I said, "I'm looking to buy an old cottage". "Well", said he, "I know where there's a couple of derelict cottages not a stone's throw from here, but, you'll have to find out from the farmer whether he'll want to sell them". "His son's getting married soon", he said, "and he's just bought him a farm, maybe the old man could do with the money"? "Where are these cottages"? I asked. "Well", he said "go along this lane until you can't go any further, then turn left and they are in the dip on the right hand side of the lane, you cant miss 'em". "Oh thank you very much", I replied, and pushed off as quickly as I could.

Chapter 38

COTTAGES IN THE DIP

As we approached the cottages we were completely overwhelmed. In appearance they had everything that we had been looking for, but they were very primitive. Their beautiful mellow brick, and sandstone mullions with Coalbrookdale cast iron lattice window frames were a very attractive feature and a characteristic of this area. The property lay well back from the lane, and the front boundary wall was built of sandstone. Although the cottages were derelict, and had been empty for many years they had a great potential. There were two dormer windows, which were in good condition, and a front porch that was central to the building; inside the porch there were two entrance doors, one for each cottage. This was an advantage to us, because we could make it into a single cottage with virtually no alterations having to be made to the outside of the property.

Ann and I were talking as if we had already bought the place, and we hadn't looked inside, or seen the owner yet. So off we went to see the farmer. The farmhouse was half a mile down the lane. It was getting late in the afternoon and when we arrived there, the farmer was very hospitable, towards us and was quite taken up with our daughter Carol who was eight years old. We discussed the possibility of buying the cottages over a cup of tea. He was quite happy to sell them to us, and said, why don't you go back and have a good look around the inside before it gets dark, and then come again and make me an offer. So off we went, the place was open to the elements, we went inside.

The swallows had come home to breed and nested comfortably between the oak beams in the bedrooms frequently swooping in and out of the broken lattice windows and doorways. There was no mains electricity here, the water supply had been drawn from a borehole in the garden, by a pump installed in each kitchen. Candles, oil lamps and wood fires were the only form of lighting and heating. Sanitation was at the rear of the property and consisted of one old brick loo to each cottage, there were wooden seats, which in those days were kept well scrubbed; there was a space under each seat for a bucket of which the contents probably made good fertilizer for the garden vegetables.

Although the property had been empty for many years, suddenly life was beginning to stir in the old place. As evening approached rats and mice began to scurry about their business, "come on Ann let's go", I said, "we've seen enough, its getting dark and we are disturbing the wildlife". As we left we saw a couple of badgers in the garden. Bats were beginning to make their entrance into the semi-darkness from under the roof tiles. Owls would hoot, a curlew called and crickets chirped within the sandstone walls.

There was no doubt that this was going to be our home if we could afford it. After a couple of days we finally decided that we would make an offer of one thousand five hundred pounds. Of course this was refused and the owner asked for two thousand five hundred pounds. It was acceptable to us, because he agreed to install mains water to the site, which was to be included in the price. Anyhow all went well, we clinched the deal and bought the property.

The mains water was to be brought to the cottages from an old derelict schoolhouse half a mile away, a completion date was set for Sept 1966, everybody was happy including the owner who reaped the benefit by claiming his government grant.

Within a week we were back there and began to clean up the site, when who should go by, none other than Stan, the bird watcher, the man who told me about the property. "Hello there", I shouted, "How are you"? "Oh you've bought the cottages then, good for you". "Yes" I replied as I pulled some bottles of beer out of my car. "Here have a drink on me, and thanks again". We became quite friendly and got to know each other over the years.

During the summer of 1965 we worked very hard especially at weekends because our cottages had to be made liveable as quickly as possible and weekends were the only times available to us. Work at the factory was reasonably OK, and the bank orders from Worralls were secure and a boost to our business. The Samuel Withers Group of Company's had drifted into decline since the death of their director Dennis Withers; orders for locks dwindled and the Group was finally put into voluntary liquidation. This was quite a blow to us at the time because we were going to lose our cash flow from the work which I had developed some years back. Although when it came to the crunch the Withers closure did actually help me to make a very important decision in my life, which I never lived to regret.

It was now well into the summer of 1965 and we were enjoying our weekend country life at the cottages. Archie Palk, young George, and his brother John all came out to help in the great tidy-up process. The swallows were on the second clutch and there were fourteen occupied nests on the site. Some Friday nights and most Saturday nights over the next twelve months we stayed at the farmhouse which was a great help and enabled us to get on with our work at the cottages. On many occasions we were invited to go back to the farm for our Sunday lunch. This was a kind gesture made by the farmer and his wife, Mr. & Mrs. Swinnerton, and it helped us to adjust ourselves within the farming community with whom we were coming to live. We were careful and were very lucky to be accepted so quickly into the family and their friends. We had come to the country to adjust, and not to take over. The farm was known as Shellmore House and there were five sons and one daughter, all of whom became life-long friends.

Now although at this time our contract wasn't signed, we were able to do whatever we liked down at the cottages, and invited our parents to come and see what we had bought. Father was sixty-eight now and still working. He and mother used to come, occasionally for a Sunday afternoon. I remember him saying, "I think you've taken on more than you can chew here". His usual comments were always expected.

These remarks didn't rankle me anymore. We had buried the hatchet and although the old chap never mellowed or backed down in any argument I respected him more at this time than ever before. It was just great to see them both turn up and take an interest in what we were doing.

Shelmore House "The Swinnertons".

Ann's mother and father used to come as well. Her mother was always worried about us, "living in the back or beyond", she would say. They all seemed to put the damper on us but in a nice way. I think that they thought that we had taken on too much work considering that we couldn't afford to pay contractors to do any work for us.

Now that the mains water situation was underway we had to get cracking on some sort of power and lighting, so I contacted the Midlands Electricity Board and asked for an estimate, to bring power to the site. They quoted the following: "two thousand pounds total costs to be paid in full in advance". It was to be agreed that work would not begin until eighteen months after the payment had been made, also we had to guarantee to use a certain amount of electricity or pay for it, and the amount set was quite astronomical. This was ludicrous and quite impossible. Our next consideration was to buy a generator. Through a friend we had the property professionally wired in exactly the same way as we would for a private house mains service. The electrician who carried out the work also found us a generator in Penrith; it was advertised for sale at three hundred pounds and was manufactured by R. A. Lister of Dursley; it weighed about one and a quarter tons, output 11KVA, 50C, 230-250 volts AC. I went to Penrith to collect it travelling through the night and returned to the cottage the next morning where I was helped by the lads from the farm to offload it and sheet it up whilst we prepared an out building to house it in.

We estimated that we would need about two tons of dry mix, (sand gravel and cement) to make a suitable bed for our machine, so we dug a pit in the outhouse accordingly. We tried to get a load of ready mix, but it was out of the question. Fortunately there was a builders yard in the village of Gnosall: A. Walker and Sons undertakers, carpenters and hardware, the usual country village business that sells everything from a coffin to a bag of sand. He was ready to oblige and was willing to help us load our materials onto the lorry if and when required.

We returned to the cottages to sum up the job. A hundred yards down the lane from our property, there was a brook running under a small bridge. We needed the water from the brook to mix our concrete. So Archie Palk and I agreed that on a day when the brook was high we would go to Walkers in the evening, collect our sand, gravel and cement (2 tons dry). We would then return to the site, park the lorry on the side verge and dry mix our materials. Then we'd take it in turns to go down the bank to the brook bringing back two buckets at a time until we had our concrete mixed thoroughly, after which we would drive as quickly as we could to the rear of the cottages, shovel it into the pit that we had prepared. It was six hours hard work, non-stop! Two days later our machine was installed and ready for use.

Chapter 39

PUT-PUT-PUT, THE SOUND OF THE DIESEL

It was very exciting when we first struck up our diesel twin cylinder, put-put-put, as she struggled to let a little air out of her system, until finally she levelled out as sweet as a nut put-put-put-put-put-put-put-put-put. Listening to the sound of our engine was similar to that which I had heard on the canal boats; it was a sound that we lived with through our future years, and not unpleasant when one had got used to it.

We had a little bit of a party, amongst the bricks and rubble, a few drinks, with our friends, and some grub to celebrate the introduction of electricity to the cottages for the first time ever. It was great. We were over the moon, the simple things in life were important especially to us.

By October, 1965 the swallows had left and were on their way to a warmer land, they had reared three clutches that year. The Autumn was an important time for us as we were busy trying to get the property secure for the winter months ahead, because we knew that we would be restricted for time and travel during the cold spell. Archie was busy glazing windows, with small diamond panes of glass that had been cut to various patterns. Poor old Archie was fed up. He had two hundred and forty pieces of glass to put in, and it took sometime to do, because he had to clean the old putty out of the iron frames. It took us almost up to Christmas to get the place reasonably secure. Our weekends during 1965 were hard work, but very rewarding.

Chapter 40

JANUARY, FEBRUARY, MARCH

For the first three months during 1966 I was heavily committed at the factory. We were very busy delivering and installing safes, there was also a fair amount of business involving safe removals coming our way. On many occasions we were short of casual labour, and so I had a talk with Farmer Swinnerton to see if his lads could help us, I knew this time of the year was their slackest time and I also knew that old Swinny could be a bit awkward, but it was worth a try. "Well ah-er-um," he muttered as was usual when he seemed to be most uncertain of what he was going to say. "I suppose it's OK as long as it dunna interfere with work on the farm". "Are you sure"? I asked, persistently, not wanting to upset him. "Well, urm, yes", he replied, "we anna too busy just now, so if yer need 'em ring me, but remember, you conna 'ave 'em in the summer, because I need all the men I can get for the harvest, in fact I might be askin' you to return the favour". It wasn't long after asking, that I needed some help, and the two-oldest sons, Robert and Edgar were only too pleased, and they came over to Wednesfield in the farm pick up, which we were able to use. This was an added bonus to us, because it was ideal for humping small safes about.

After our first day's work with the lads they became more eager to come again especially after they had been paid a few bob. Robert, the second oldest son, always liked to come with me on jobs, safe opening, removals etc., and he was quite a good learner. We were exceptionally good friends. Although he was only twenty two, he was a good farmer. He was also very keen to learn other things in life; he had a hankering to want to know more about the outside world and was much more outgoing than his brothers who lived a secluded life on the farm they nicknamed The Ponderosa.

Now getting back to the factory. Work inside was quite hectic and the outwork on the bank contract with G. Worrall & Sons was increasing. Sometimes there were attempted break-ins, loss of keys, and general wear and tear of locking systems. This was now building up into a regular service programme with the bank.

"FRUSTRATED TRAVELLER"

Highways Byways. Travelling fast and far,
    Out all hours in my motorcar.
A frustrated traveller, who needs to use the phone,
    To give a message to a customer, and make a call to home,
There's a little red box, this will do me fine,
    Until I stop, and then I see the old familiar sign.
'No Waiting', says the warden. 'No waiting here today',
    So I jumped into my motor, and quickly drove away.

Off I went in my motorcar, travelling fast, travelling far,
    I found another phone box in a quiet spot,
But couldn't get my money in the ten pence slot.
    Thwarted now, my nerves are getting frayed
Two telephones I tried, and not one call I made.
    Off I went in my motorcar, travelling fast, travelling far,
Determined, I will try again,
    I pulled up at a call box in a country lane.
Picked up the receiver to make my call to home,
    But alas, I was foiled. There was no dialling tone.

Off I went in my motorcar, travelling fast, travelling far,
    Again I stopped to make my calls, as quick as I was able,
Seems I'm out of luck today, someone's cut the cable.

Happy just now, not looking for a phone,
    My customers closed and I'm nearly home.

George and I were travelling the country covering Tees-side, Tyneside, Lancashire, Yorkshire, North and South Wales, and the Midlands as far south a Northampton. There were also the Police Stations, Law Courts, and other general callouts for Lingham Bros. of Birmingham. Things were going too fast and it was obvious that I had to do something about it, because I could not control the out work and the factory. In addition, Spring was just around the corner, and it was time for us to return to the cottages and the countryside that we loved so much.

Chapter 41

HARD GRAFT AHEAD

We returned to the cottages during late March of 1966 to hear the call of the curlew, also the lapwings created great attraction as they swooped overhead whilst others sat comfortably on their sparsely built nests in the ploughed field behind our boundary. I remember when Ann and I made flags so that we could use them for marking out where the sitting birds were; this was a great help to the farmer especially when spraying the wheat. Another welcome sight came on the twelfth of April: we sighted the return of our first swallow. I called out, "open the windows, the swallows are back" and so the windows were thrust open.

After a discussion we agreed to leave one bedroom window half open for the whole season and also the windows in the out buildings could be opened every Spring, if and when the birds returned.

After a couple of days there were more swallows drifting in. They were soon busy repairing old nests or building new ones just as their ancestors had done over the past years.

We were very surprised that they came back to rear their babies in what was now our Generator House. It was quite noisy in there with the sound of the diesel engine, whereas up until a few months ago it had been perfectly quiet, but they loved it and thrived there, and they still come here today thirty years on.

We have estimated that since we bought the property in 1965 to date 1997, somewhere around three to four thousand swallows have been reared here.

I remember coming to the cottage one summer's evening. I was about one hundred yards from my gate when a swallow hit the front of my car, I looked through my mirror, its wings flapped erratically whilst it lay helpless in the centre of the lane. I left my car at the gate, and ran back as quickly as I could but I could feel no life there. I put it in my warm cupped hands and went back to the cottage. I must have nursed it a long time perhaps ten minutes, it seemed like hours, then just as I was going to give it up as dead, I felt a very slight movement, "its alive Ann, it's alive". I kept my hands cupped together to give warmth and darkness until it regained consciousness, after which I went outside, opened my hands, and there it sat in the cup of my left hand as snug as a bug in a rug. It stayed there for about ten seconds before taking off on the wing. It had been a happy ending and I was very pleased with my first attempt of country life saving.

For the next few months it was hard graft ahead. We were working weekends flat out and we had now made access through to each cottage. Although we still had two flights of stairs and two kitchens, it was now one cottage.

There were six small bedrooms, four of which we were converting into two large ones, leaving one small bedroom to the swallows for that season, This bedroom was eventually to become our bathroom, but, we were in no hurry. There were more important things to do, for instance the other small bedroom was to be for our daughter Carol. We quickly made this into a very pretty room for her, so that we could move in by the middle of July at the latest, giving her time to adjust and to get to know children at the farms around before starting her new school in the Staffordshire village of Gnosall.

However this was not possible because we also had an agreement with Post Office telephones that our service would be connected by the fifth of September 1966. This was very important to us for the business and other communication. The total cost for the installation was to be ten pounds which included all labour and materials, and the setting up of thirty two telegraph poles. This was an excellent service carried out by the Post Office and was completed on time at the agreed price.

During the late spring of 1966, I thought it might be a good idea for us to have a puppy and start to train it before we moved in our new home. Ann was not particularly in agreement because she knew that if we bought one whilst staying at my in-laws - knowing how fond they were of dogs - they probably would not have wanted us to take it away when we left, so I went along with that, until a few weeks later one day during the middle of June, 1966 when I was in a pub in Willenhall having a few drinks, I met a mate whom I hadn't seen for a long time. He had a puppy with him, it was a beautiful Alsatian and it was only about twelve weeks old.

We got talking over our beers and I admired the pup. "Do you want her?" he said, as he threw her a couple of small pork pies. "She has a bostin pedigree" he said, "she belongs to a friend of mine, he lives in a council flat. He's been smuggling her into his flat at night and I have been having her during the daytime, but the Council has rumbled it, so she's to be put down if I don't find her a home within twenty four hours, or me mate's out of his flat".

I took the opportunity and said, "come on let's have her", I could have had her for nothing, but I gave him fourteen pounds for her and it was the best fourteen pounds I have ever spent. I collected the pedigree the next day.

On my arrival home I was a bit concerned on what I had done but I knew it would be OK, so when I got home I opened the door and bundled her in. They were over the moon, I remember father-in-law saying, "if you don't want her we'll have her, but after Ann had seen her there was no chance of that. She settled down straight away in a box her big plodding paws lopped over the box edge, and she never looked back. She didn't have a name, so we called her Mandy.

Carol was now nine years old. She had made good friends with kids from neighbouring farms, and especially the children from the cottage a quarter of a mile away from us. They had six kids, four girls, two boys. The father was known to everyone as the local poacher, who, for the sake of the story, I have named Joe. Pheasant, partridge, ducks and hare, Joe looked after us so well I thought a small poem wouldn't come a miss:

POACHER

He's off to the woods, this man I know.
The poacher, he's known as local Joe.
Now this man Joe, has never been caught,
He's always used his head,
For he only went out in the early morn.
Whilst the keepers were in bed.
He'd fire one shot,
A bird he'd got.
Then Joe would have to flee.
Before he went, he'd light a fag,
And wedge it in a tree,
He'd stand around.
Wait a while.
Knew what tricks to play,
By the time the keepers came,
Joe had slunk away.
At first, they'd smell the smoke,
Then they'd see the fag.
In the distance, another shot,
That bird was in the bag.

Now Joe's cottage was a rented semi with a big garden. The loos were outside at the back were made of wood with a bucket under, the contents of which were emptied at the bottom of the garden in a pithole.

The adjacent cottage was identical although it was now in poor condition and had not been lived in for many years.

The only water available was drawn from a borehole to a pump in Joe's kitchen. Water was heated by fire only, and Joe's wife cooked everything on that fire. She was a good, hard working, Irish woman and one wondered how she brought up six children in those days under such conditions. Unfortunately, she is dead now but the six children are a credit to her.

Lighting in their cottage was only by paraffin lamps or candles. Baths became very difficult for them, especially when the kids got older. The girls, being good friends of Carol's, would sometimes come to us.

Joe and I became good neighbours drinking frequently at the Red Lion, which became my local pub for many years where I played darts and listened attentively to the stories of keepers and poachers of which Joe was the latter.

The poacher who never got caught, 'Joe', although in later years he became a game-keeper.

Chapter 42

STAYING AT THE FARM

During the summer of 1966 we found that staying for weekends a farm was becoming more like a holiday. Over the previous months we had really enjoyed a touch of farm life, but it was now loosing us valuable working time at the cottage. I had helped Robert with the milking on Saturday and Sunday mornings. Ann and Carol washed out churns and units, other times we fed calves, looked after hens, and helped catch sheep for shearing.

One day I remember walking across the fields with Farmer Swinnerton when he stopped and pointed to the sheep from a distance and said, "look ya, that sheep's got ticks, I shall av ta dip them this week". "I can't see the sheep, let alone them ticks", I replied. It had been a lot of fun staying weekends at Shelmore Farm, Ann and Carol had slept in the same bedroom as Mrs. Swinnerton's daughter, Barbara, whilst Archie Palk and I pitched in with the lads at the top of the house. Archie and I would generally bunk down on our camp beds at about twelve o'clock; the lads came in at all hours usually half cut, each one tipping us out of our beds so our sleep was at a minimum.

I remember one night in particular when the lads were locked out, we were awakened to here a scrambling up the roof tiles and a banging on the window "come on open up, they've locked us out", they cried. We opened the window and hauled them in. By the time we got down to kip again it didn't seem five minutes before we were up for milking.

We had got to know the Swinnerton family very well and they provided us with a good insight for our future country life. It was now the beginning of September 1966, our cottage was just about liveable, the telephone service was connected, a piped water supply was laid on, and our generator was working well. So we were ready to leave Wolverhampton and begin our country life proper.

Chapter 43

MOVING OUT, MOVING IN

On the third of September, 1966 I left my in-laws house to fetch our lorry from the factory, and then returning there to load our furniture. It was all go and all aboard, Ann, Carol, Mandy and me climbed into the cab and off we shot.

My in-laws were sorry to see us go, They thought that moving into Shropshire was the end of the world. In fact as we left mother-in-law shouted, "don't forget to get plenty of provisions in stock its lonely out there, and there are no shops".

We took the A449 road in the direction of Stafford, and we hadn't gone ten miles before the back tyre went pop. I got out of the cab looked for the jack, it wasn't there. After a bit of chuntering and cursing to myself, I realised that as usual I had to improvise. Whilst I was changing the wheel Ann, and Carol stood by on the pavement and watched.

Looking at the furniture sticking up on the back of the truck we looked and felt like a lorry load of didicoys making their way to a site. Eventually we arrived at our cottage, humped the furniture inside, put it amongst the bricks and rubble and sheeted it up. When I put the lorry away for the night I thought it quite touching to find a large carton behind the seat, mother-in-law had packed it with tinned foods for our stores. Our first night was very strange; the sounds of the wildlife were tremendous just as if they had come to welcome someone back into the old place. Carol slept like a bug in a rug, and so did Ann and I until we were awakened by the call of the cowman – wut-wut wut-wut. Almost a week had passed and Carol was about to start her new school. The school bus was to pick her up at the cottage in the morning and bring her back in the afternoon; this was very reassuring to us to know that our daughter's safety was in good hands.

Chapter 44

A JOB IN COUNTY CORK

It was October 1966, we had been living in our cottage about a month when I received a telephone call asking me to quote for a job on board ship at the docks at Cork in Ireland. My fee was ninety pounds, and if I were going to make a fair profit then I had to be back home on that same day, or the next day at the latest.

I was given the OK to go, I made my arrangements and travelled to Birmingham Airport where I garaged my car. After making enquiries at the desk I was told that there were long delays in flights to Ireland, so I pursued the possibilities of flying to London, and from London to Cobh near Cork. They were favourable and I was able to make a booking alteration and away I went. On arrival at Heathrow I was in luck. The time schedule was OK, and the plane took off for Cork. Somewhere on route we were diverted to Shannon Airport because of thick fog, and then we had to wait for taxis to take us to our destination.

I arrived at Cork somewhere about six o'clock in the evening; I found a small hotel there and started my work the next day. By mid-day I had finished the job and then the fun began. I returned to the Hotel, paid my bill and went to the Airport in the hope of getting a flight home. It was obvious there was no chance. Fog had persisted since I arrived there, so I went back to the Hotel where I had to stay for a few more days, during which time I mixed with the locals, and drank away all of the profits. Late on the fourth day the fog had lifted considerably, and the possibility of a take-off to Birmingham looked encouraging. I went into the Airport lounge, supped a couple of brandys whilst I waited and then boarded the plane. There were only about eight passengers and there were plenty of seats available. So I sat alone until I was joined by a big burly Irishman who introduced himself as Pat Murphy.

Obviously the bloke wanted to be friendly and we soon got into conversation well before take off. We finally became airborne and it wasn't long before refreshments were available. "So", said the burly man, "What'll yer be arvin sun"? "I'll have a brandy if I may", I replied. He beckoned to the stewardess, "Two double brandies if yer plaise". He downed his as soon as they came, and was looking for the next one. During the flight he asked, "where'll yer be going"? "Stafford", I replied, "and I pick up my car at the airport garage". "Is it OK if I ride wit yer"? "Yes" I said, in a slurred voice as the drink began to take effect. We finally landed at Elmdon where I almost fell down the gangplank. Pat was OK, and none the worse for booze he followed me to the garage and we were soon on our way. He hardly spoke a word to me between Birmingham and Stafford until I dropped him off at the railway station, where he got out of the car double quick and said, "Tank yer kindly sir, but, oi'l not be riding wit yer again, yer too bloody fast for moi loiking", and away he went.

When I arrived home I totted up my expenses from the job, which had resulted in a loss of seventy pounds. Who cares anyway? To hell with profit, I'd had a bloody good time.

Chapter 45

BACK AT THE FACTORY AGAIN

I was very unsettled during the year of 1967, I also became very agitated because I knew that I didn't want to leave the cottage every day to come into the Industrial Midlands which was now becoming a rat race and a major problem to me, because my love for my family, our cottage, and the countryside was beginning to play a larger part in my life than I would have ever imagined. I could see and feel the business gradually slipping into fourth place.

There were many occasions when I would arrive at the factory in a morning, only to find that emergency jobs were waiting for me, and at this time most were in the North of the country, in the direction and beyond where I had gone to live. I had travelled twenty miles south only to return in the same direction from which I had come.

The area we were covering was now expanding rapidly and we were the only service engineers outside the city of London who were available to carry out service work for the bank in connection with their refurbishment programme at that time. It was around the clock service, seven days a week with immediate call out commitments in most cases. Young George was happy travelling and coped admirably with the North and Mid Wales side of the business.

Uncle Tom of Worralls Locksmiths London gave us our bank work, and as long as we looked after it we would never be short of a bob or two.

However as I have previously said, throughout my working life cut throat prices were always a bugbear, the trade as a whole was not of the best when it came to payment for labour, or lock sales. In fact most traders who wished to purchase our locks would often state how much they were willing to pay for them. We would have to get down to that price if we wished to be favoured with the orders.

In some instances Worralls were no exception, especially relating to service work which they paid at a rate of fifteen shillings per hour and six pence a mile (240 pence to £1.00 1965/67). At this time our work force was thirteen, which included four women who operated the fly presses and assembled locks, and one power press operator.

We also had three skilled locksmiths apart from George and myself, and two part time male employees were also kept on the books to carry out safe removals and installations. Last but not least, Maureen ran the office and was a great asset to the business.

Father would come to see us quite frequently, and his visits were most welcome. He was still working at the oldest firm of locksmiths in England James Gibbons in Wolverhampton and his skills were greatly valued by them.

During Christmas 1967 he was taken ill with a heart problem, and I always remember him saying to me "Bob, I never wish to see another January, I 'ate January, always 'ave done". He died five days later, on the evening of the 31st December, aged sixty nine.

It seemed ironic that his wish came true, and I have often wondered whether serving in France during 1914-18 had left him with bad memories of the winter time which he had talked about when we were younger.

During 1968, mother decided that she would make a move from the little village of Wombourne in Staffordshire. She was set on coming to live nearer to us, so she bought a bungalow at Newport in Shropshire. However, it wasn't long before she became restless and wanted to move again. This time brother Geoff and I got together and helped her to buy a cottage in the village of Gnosall about two miles from us.

After we had spent some considerable time helping her to settle down again we were able to continue with our work.

Ann and I had grafted hard at Guild Cottage, and it was coming together very well, we had thought of changing the name of the property to something like Badgers Covert, but after much thought we decided to leave it, as it was original and named after the Guild 'O' Monks.

The alterations we made there were considerable but generally we had not modernised it, although we had taken out fifty-two feet of walling inside we didn't alter the character in any way.

Ann and I were a great team. Many times when I had been called out on jobs I would return home to find that a wall had been removed, or some oak beams exposed. We never had to tell each other what had to be done, it was always very automatic for either one of us to sort out the work. In fact we were so happy working together that I was beginning to think how great it would be if I could make locks and do the service work from home instead of travelling back and forth to Wednesfield.

I could spend more time in the garden, and there was a marvellous country life to go with it.

Back in Wednesfield our work for the bank was building up nicely especially the service work. The more the build up, the more I saw the potential of working from home until finally I began to think seriously about selling the factory.

At first it seemed ludicrous that I should even think of such a thing. The old fella would have turned in his grave at the thought of it. I can imagine what his comments would have been, "Oh dear wots' 'ee bloody well gone and done now".

During the following twelve months I had made my decision, I was going to sell. It was not easy, I had employees to think about especially young George. He was a good chap, loyal and dedicated to his work and of course I would give him good references.

I wasn't so worried about the others. There were plenty of jobs about for good workmen. Then of course there was the casual labour, well they were only after beer money and they also had other jobs elsewhere. After a lot of thought I came to the conclusion that it was about time that I considered my family and myself.

So my mind was made up, the factory was going to be sold and the cream of the work was in future to be made in a small washhouse at the back of our cottage. George managed to get a local job but he wasn't very happy. Anyway after about three months I received a letter from Chubb Locks, London, asking for a reference on George. I was very pleased that I could help him. So I wrote out a reference and filled in the request, he got the job and is still there today almost thirty years on. George is now fifty-six and I am seventy-two.

Its great to think that he still comes to see me, the son of his old gaffer, who gave him his first job from school at fifteen.

There was no doubt that I was going to be very busy making rackbolt locks for the bank, and travelling the country doing their service work, but I had thought it out and I was quite happy to get clear of the Industrial Midlands and work from home. I soon found a buyer for the premises and I was ready to move out.

I also sorted all other outstanding business as quickly as I could. I had discussed my future plans with Uncle Tom of Worralls Locksmiths and they were obviously very pleased because it meant that I was more dedicated to their business with the bank than I had been previously.

However I knew my business and if I looked after it properly my living would be secure for the rest of my working life. I also knew that there was no fortune to be made by working my trade and business from an old washhouse at the back of a country cottage, but I wasn't looking for a fortune, I was looking for a good life and it was staring me in the face just waiting to be snatched, so I grabbed it. I took the machinery and press tools I needed and went back to the cottage.

The machinery consisted of a driller, a linisher, a lathe, which I had converted into a key cutting machine and a polishing spindle.

By having just enough machinery and tools in the wash house this would help me to work more comfortably within the small space that was available to me.

I had supplied Worralls with a stock of rackbolt locks which I knew would last them about three months. During that time and in between going out on jobs, Ann and I knocked out the old boilers and we renovated the washhouses. Although they were very small, they were respectable and looked good. We installed a workbench, ubbin, and leg vice, and then I was ready to start making locks at home by hand.

My small amount of machinery was driven by single-phase motors, which matched our generator. After using it for about three years we had a major breakdown and it was scrapped.

Fortunately, I happened to know someone who would sell me another machine, it was an old Petter Diesel made in about 1937. It had been used for the rearing of pheasants at a country house "Loynton Hall" near Newport, Shropshire. I bought it for twenty-five pounds from the lady who lived there. Although it was old it was in good condition. I installed it in the generator house at the cottage, and I worked it hard for twenty-five years, from 1970 to 1995.

One would hardly believe that all the rackbolt locks made for the Midland Bank for their front doors during that period, were made at the rear of our cottage in converted washhouses, of which the electrics were produced by an old water cooled diesel engine driving a six KVA auto start generator which cost twenty five pounds!

We also had a 1.75 kw RA Lister Diesel air cooled SLI start-o-matic generator which provided our cottage with lighting and television for thirty-one years which I purchased for fifty pounds.

Chapter 46

ON THE CHEAP

During the early part of 1969 Ann and I were going to knock the two kitchens into one at the cottage, and I was on the look out for an R.S.J. girder on the cheap.

I approached young Bob Swinnerton to see if he could help. "Yes", he said, "give me the size of the girder and I'll have a word around", "OK Bob, see what you can do for me". Within the hour Bob was back on the phone, "I think I've got what you want, and it'll only cost you a couple of quid". "OK just the job how do I fetch it" I asked, "We'll fetch it in the morning", he said, and so the next morning Bob hooked his trailer to the back of his car and came to the cottage at ten thirty. "Where have we got to go", I asked. "Oh" said Bob "its only at Newport, but we have to pick up a bloke at Sutton".

Sutton is a small hamlet about two miles from our cottage, and halfway to Newport. We stopped to pick up this bloke, his name was Jim and he was a Woodman cum Dealer, and Bob had negotiated with him for our R.S.J.

Just a few yards from Jim's cottage was our local pub called "The Red Lion". Most of the cronies there were Jim's mates. Jim suggested that we go in for a jar, so I bought the drinks, brandys for Jim and beers for Bob and me. After about an hour we moved on towards Newport. We hadn't gone a mile before Jim wanted to pull up at another pub, "The Swan at Forton". "We'll just pop in for a quick un". Says Jim. I asked "what about my girder"? "Oh dunna worry about that, it inner gonna goo anywhere". Then he made his way to the bar and said, "Mines another brandy". It was obvious who was paying today, me, Billy Muggins again. After a few drinks in the Swan I was beginning to enjoy the outing, and in any case Jim wasn't going to take me to see the girder until he was good and ready.

We reached Newport at about twelve thirty and Jim insisted that we should pull in at the Barley Mow, his wife worked there, and he wanted to see her.

All three of us went into the pub Jim bought his own drink and borrowed a fiver from his wife and off he went into the dining room for some grub. We followed him sat down, and decided that we had to eat too.

So we all had a good feed and I was again the mug with the money. Finally we reached our destination. I had already given Jim a couple of quid for the girder, which he probably got for nothing. However it was just what I wanted, so we loaded it onto the trailer and by the time we left Newport it was almost six o'clock. The pubs had reopened, although to us they hadn't been shut. So we ended up back where we started The Red Lion at Sutton. Jim was going to pay this time but he didn't know the meaning of the word when he was half cut.

He was quite a character and although it seemed at the time to be a most expensive way of buying a girder, it gave us further initiation into our community and country life. Although Jim was a drinker, and I had spent a lot of money, it could be said that buying on the cheap becomes the most expensive, but, in the long run knowing Jim during our day's outing paid off and we became good mates for many years.

At this stage we were planning a layout for our kitchen, the rough work involved was to knock out the dividing wall and remove the old baking ovens, which were not practical for us to use. This meant that the finished size of our kitchen was going to be about twenty feet long by twelve feet deep, and we had been offered an almost new four-oven AGA unit free of charge. All we had to do was dismantle and collect it, so off we went to Stourbridge with the lorry to collect what was to be our prized possession in cooking equipment. Ann was very excited about it and anxious to get it working, so much so that the day I had set up the ackrow jacks and put in the RSJ. Whilst I was called out on a long distance job she was taking down the wall and removing the baking ovens. In fact, when I returned home the job was done and the place was cleaned up and it wasn't long before our kitchen was prepared and our AGA installed ready for use.

During the summer we had bought hen-houses and duck-pens and were building runs for them. We also had pigstys and at the time we were rearing our first pigs, three in all, one for curing our own hams and bacon, the other two we would sell for profit if possible. In our pantry there was a settle just waiting to be used again, so when the pigs were ready to be killed we would send them to a small abattoir in Market Drayton. The owner's name was Geoff Farr. He was a butcher and he had a shop in Market Drayton and a market stall.

Our good friend Farmer Swinnerton collected our pigs for us and delivered them to the abattoir where Mr. Farr would kill them and return one to us for curing. This was a good system and worked well for many years. Ann was very distressed over this because she had a passion and love for all animals. She had grown very fond of them, and didn't want to see the dead one come home. I would get about twenty eight pounds of salt, boxes of saltpetre and some pepper ready for Geoff Farr for when he brought the pig back. Then he would show me how it was done. It was salted down on the settle, rubbed in well, flitches, hams, and loose pork of which the latter included chops which were eaten during the six weeks of curing. Also there was the lard and an added bonus of tasty scratchings and home-made Brawn from the pig's head.

After the curing process we would pick a dry day when we would take the hams and flitches and wash off the salt in the bath, then hang them on the garden fence to dry. They were then peppered on the bones wrapped in butter muslin hung on the hooks on the dining room ceiling ready for use, and that was our breakfast for the next twelve months.

Whilst I'm on the subject of pigs there was a day when a local farmer telephoned me to say that a relative who was a Reverend was visiting them. He had arrived in London from South Africa and he was coming to visit and stay with them for a week or two, but, the purpose of the call was that he had lost some keys which operated three Chubb deed boxes. I was asked if I would open them and cut new keys for him. This was to be another nothing job because I knew that the people who telephoned me were as they say, as tight as a drum.

Although they had plenty of money, whenever they came to pay me for anything they only had a ten bob note in their pockets. So the Reverend came to see me with his boxes, and introduced himself as the Reverend Tyler Wittle. "Please will you open these boxes for me and make new keys, and could I collect them next week" he said. "Yes OK", I replied, "leave them with me, I'll telephone the farm when they are done. A week later he came to collect them. At the time we were cutting some ham for our lunch when he said, oh that looks good dare I ask you for a slice. Yes its home cured I replied and I gave him a couple of slices. "Ah thank you my son", he said as he tucked it through a slot in his frock and put it into his inner pocket and left. A few days later there was a knock at the front door, it was the Reverend, "Hello", I said, "are your locks OK"? "Yes thank you very much", and he continued, "you know I did enjoy the ham you gave me last week it was beautiful, do you think I could have some more"?

Well I thought that takes the biscuit, I had done the locks for nothing, given him my ham, and now, like Oliver Twist, he asked for more.

Chapter 47

FOOT AND MOUTH DISEASE

The great tragedy for farm animals and a disaster for farmers is the infection of foot and mouth disease.

Farms in Staffordshire and Shropshire were all affected in one way or another during 1967-68. Hundreds of cows, sheep, pigs were being slaughtered and cremated within fields of pasture where they had once grazed.

As I passed through the county of Shropshire on my way to the Midland Bank at Oswestry I could smell the burning flesh of animals, I could also see the black smoke on the horizon and I could see the diggers working non-stop over the troubled farmland.

Police guarded many premises confirmed with foot and mouth infection. Notices were displayed in many parts of the counties of Shropshire and Staffordshire. On my return journey home I was to learn that the Ministry of Agriculture inspectors had been to visit all farms in our area. When I arrived home Ann told me that they had been to inspect our pigs. Recently we had acquired a couple of goats and they all had to be checked out by the Ministry.

The inspectors wore protective suites, gloves and Wellingtons and were disinfected with Jeyes Fluid before entering and leaving our premises. Our small amount of stock was cleared. The movement of animals was for some time under strict supervision and by permit only.

From November 1967 and continuing into 1968 there were one hundred and thirty four thousand animals slaughtered as a result of foot and mouth disease, the highest number since 1923 when seventy three thousand, five hundred animals were slaughtered.

Chapter 48

THE RED LION

Most weekends I used to go to the pub at Sutton. It was two miles from our cottage. Sometimes I would pick up Bill the poacher. Bill had a bike and after a skinful he had been known to tumble off and end up in the ditch where he would stay until he sobered up a bit.

We would play darts until all hours. The landlord was known as Fergie and he was quite a character. When he wanted to go to bed he would say "if you want anymore drinks serve yourselves, put the cash in the till, and drop the latch on your way out, Goodnight." He kept two Labradors and a muscovey duck. His duck would walk up a plank to the point of balance and topple over into a tub of water, until one Christmas it was stolen, and never seen again. Poor Fergie was devastated and always wondered whether it had ended up on someone's table.

His dogs were so well trained they would collect the empty pint pots by the handles when instructed by Fergie. Also if the dogs were given a sixpence by the customer one dog would hold it in his front teeth and both would trot off to Fergie. Fergie took the money and gave the dogs a bar of chocolate. Both dogs would return to the generous customer and he would share the goodies between them.

I always used to take Joe the poacher back home from the pub. I remember one night when I dropped him off, he said, "come up in the morning, and I'll give yer a couple of brace of pheasant". So I went the next day, the kitchen was full of pheasants hanging from the ceiling; there must have been at least thirty brace apart from ducks and partridge etc. They all had come from local breeding stock, not that it bothered me I was only too happy to have them.

Some years later after their children had grown up, Joe's wife Mary left and went to live in Gnosall. I used to call in and see how he was getting on because he was living by himself. I remember one day when I went through his kitchen to the lounge, being careful not to tread on the Bantams that clucked around my feet. When I got there I saw that Joe had got his black and white battery telly and I was amazed when I looked behind the door on his old couch was his goat. She sat upright like a human back legs wide open, udder full, a true British Sarnan watching the telly and waiting to be milked.

Getting back to the locks, other than Worralls of London, I was the only locksmith in the country involved in the making and servicing of rackbolt locks for the Midland Bank. It didn't take long for me to get the thing going at the cottage. Our generator was in full swing, and I soon realised that this was the lifestyle that we had been looking for. My visits to the lock country soon became infrequent. I would go just to fetch castings and other materials for my work.

I had turned my back on the Industrial Midlands for the last time.

We hadn't been at our cottage long before we became interested in hobbies such as bee keeping and gardening. Ann and I were a great team because we had found a simple reward for our labours, which we had strived to achieve. Living and being accepted into the farming community was just great.

We were to make our life the good life.

Chapter 49

A HOBBY FOR
TWENTY FIVE YEARS
– BEEKEEPING

During my frequent visits to the Red Lion, I became friendly with a chap named Reg Heathcote. Reg worked at Aqualate Hall doing property repairs on farm cottages and the Hall itself.

He had been a bee keeper for a few years and he offered me a hive of bees at a reasonable price, but I knew nothing about bees. Anyway Reg said he would help me to winter them down. "OK Reg I'll have them, can I collect them tomorrow"? "OK, Yes", he said "make it during the evening and I'll close them in ready for collection.

I fetched them the next day. I had cleared and prepared a site ready for their arrival. We put the hive on the site facing south as instructed by Reg and we kept the bees closed in for a couple of days, so that they would not return to the old site. All I had to do now was to feed them, put on the quilts and wait until spring, "I thought"!

In the early spring I watched anxiously for a sign of life from the entrance, but there was nothing. It wasn't long after this when I was visited by an inspector from the Ministry of Agriculture. He informed me that the bees had starved and that there were only about fifty workers and a queen in the hive. He left me in an awful mess. I was devastated and didn't know what to do. However two days later I read an article in the Newport Shropshire Advertiser. A beekeeper in Market Drayton was offering to help budding beekeepers. So we jumped in the car and went to see a man named Maurice Lewis. Mr. Lewis was only to pleased to help us and it was obvious when seeing his awards around his cottage that he was the man to help.

So he said if you will bring me back I'll come now, the sooner the better. When we arrived back at our cottage he looked in the hive and said starvation is the problem here. I was very upset because through inexperience I had made a bad mistake.

Beehives at the Cottage.

Maurice removed some of the combs for me to look at, the bees were dead, their tongues extended. They had tried to get food from the cells, but there was none there.

The remaining workers about ten, had cleaned out the larder to feed their queen, it was a sorry sight. Well said Maurice, "I'll take the queen, I can use her, also I will get you a six bar nucleus and you will have to start again. In the mean time I burnt all the old combs, disinfected the hive and waited for my new stock.

When Maurice came to set up the hive he gave me my first lesson. I was to handle the bees by removing a comb and stroking them lightly with the back of my hand. I had a couple of stings that hurt, but I had convinced my teacher that I would make a good beekeeper.

Before Maurice left he looked at the feeders which I had been given with the bees. They were old fashioned and too small so I bought new ones half gallon size because I never ever wanted to see bee starvation again.

Maurice left me with good advice, (if in doubt feed them). Maurice and I were good friends and I kept in touch with him until the day he died.

On his advise I bought the book "The Hive and the Honey Bee", by Dadant & Sons, a book which I have learned so much from over my twenty-five years of beekeeping.

"BEEHAVE"

Call out the Guard, there's something funny,
  The keeper, he's come, to steel our honey.
With smoker, gloves, clad in veil,
  He we hope will not prevail.
We're ready girls, our stings are sharp,
  We'll make his life a little harder.
Alert yourselves, defend your larder,
  He puffed his smoke into our home,
Took our honey, left us alone.
  We'd lost our food, and lost our fight,
No flowers, or trees to put things right.
  It is hoped that he will remember.
To feed his bees around September.
  As a keeper, bees he'll cherish,
Without his care,
  No doubt they'll perish.

Chapter 50

THE GOOD MOTOR

During the winter of 1969 an old Uncle of Ann's came to stay with us at the cottage for a few months. His name was Eddy and he was coming up to ninety years of age. He was a tough little man, a Farmer by trade. He and his wife went to Canada just after the First World War about 1920. They bought quite a lot of land there, it was as cheap as rags to buy, in fact the Canadian Government were almost giving it away, this was done to encourage prospectors to go out there and make a life for themselves.

Eddy built a log cabin, where he and his wife settled down and reared their family until they returned to England during 1937 to provide a higher education for the children.

Whilst he was staying with us he would dig the garden, do some hedge - laying along with many other odd jobs. I remember a day when Eddy was a great help to me because one night I had left my car out of the garage. There was ten degrees of frost that night and I had no antifreeze in the radiator. Two bottles of beer lay broken and were solidly frozen on the rear window ledge.

I started the car knowing that the block would be cracked, which it was, and the water began to drip as it thawed out. I had a job to do in Leeds that day, but how I was going to get there I didn't know.

Anyway I went into the cottage, and grumped at Ann. It was quite usual for me to do this when things weren't going right. Eddy heard and offered his help, "the same thing happened to me in Canada", he said. "I'll tell you what to do, got any pepper or mustard, or both"? Ann produced five tins, pepper, mustard, curry power, cinnamon and ginger. "Tip it all in", said Eddy, "try to plug the crack with something first before you put in the water, then start her up". I did exactly what Eddy had told me and after running the engine for about an hour there were no leaks, so I put five gallons of water in the boot and off I went from Gnosall to Leeds, and back home without a hitch. After which I drove the car all over the country for 11,000 miles. Finally I sold it for one hundred pounds. The new owner used it for three years, the block remained cracked until the car was sold for scrap.

I called it the good motor.

Chapter 51

MANDY

There were many times when Ann and Carol would come with me on certain jobs. Occasionally we would take our Alsatian dog Mandy with us, but only if the work I had to do was in a suitable area like the seaside.

One day I remember taking them with me to the Lake District. When I arrived there, Ann, Carol, and Mandy went off for the day. Little did I know it was going to be a lousy job. I didn't finish at the bank until eight o'clock in the evening! By the time we had eaten it was almost nine o'clock before we got on our way home. As we motored down the M6 we decided that we would come off the motorway at Knutsford Services where we could exercise the dog. I noticed that I was being followed by a police car which pulled in along side me in the parking area. Two police officers briskly jumped out and started to walk around our car, when I opened my window one of the officers said, "will you please get out of the car", the dog lay grumbling to herself on the back seat. The one officer was arrogant and abrupt, and his approach got my back up from the start, "where are you going"? he snapped. "Home" I replied. "And where's home"? "I live near Newport in Shropshire". "Then how did you come to tax your car in Stafford"? "Well", I said, "Stafford and Newport are about eleven miles apart. I live between the two towns and because I live on the border of both counties, my postal address is Newport Shropshire and I pay my rates and car tax in Stafford, do you want me to show you on the map where I live"? The next question, "Where have you been"? I promptly replied, "on a bank job". It was clear that I was winding him up, and he didn't like it. "Are you trying to be funny? I asked you a sensible question", he said, "where have you been"? "I have been on a bank job that is a sensible answer, I am a locksmith, I work with a London Company in connection with the Midland Bank, here are my credentials, now let's get on with it, do you want me for anything else"? I asked, "because we have had a big day and we want to get home". "Can we look in the car"? "Yes", I said, as the dog sprang towards the window, teeth ready for action. After which the officer said, "I'll look in the boot if you don't mind". He looked amongst my tools and then said, "OK you can go, but next time be more co-operative". To which I replied, "your approach could do with polishing up, goodnight" and then they cleared off.

So I let Mandy out of the car and afterwards we left for home.

Chapter 52

SHOOTING DAYS

Shooting Days at the Shelmores were traditional and well arranged by Farmer Swinnerton. His wife was responsible for preparing and cooking as many as six brace of pheasant for the evening meal which was always ready for the shooting party when the returned after the completion of their sporting activities.

Sometimes shoots would take place twice a week but never on a Wednesday because Wednesday was Market day at Market Drayton.

Shooting days depended on how many birds were around. Sometimes it would also depend on whether there had been any activity from Lord Lichfield's Estate which was only separated from The Swinnerton Farm by the canal and the Norbury Road.

Lichfield's Shoots were attended by V.I.P's such as the Duke of Edinburgh and Prince Charles etc., and there were thousands of birds reared each season for the occasions. So it was obvious that many birds would get away during their shoots and end up on The Swinnerton land in the Shelmore Valley only to be trapped there, and shot the next day by the Swinnerton Shoot. There was little chance of the birds escaping the kill one way or the other. I was very often invited to the Shelmore Shoots and enjoyed the freedom of roaming the countryside especially to here the crisp crunchiness of our wellies as we moved over the frozen snow into the woodlands, where the dogs became excited by the beaters who thrashed out any life that was hiding there.

I soon realised that I was not made for this kind of sport; neither did I have the instinct to kill the birds. It wasn't long before the farm lads rumbled me. "Bob never gets a bird", said Edgar, and they all agreed with each other and wanted to know why. "Can't you shoot?" said another, and then I had to admit that I didn't like the killing, and also that I had been firing at or behind the birds instead of in front of them. Eventually, I dropped out of the shooting party. I continued to go on the shoots and I volunteered to pick up the dead or injured birds.

I had been taught properly how to kill birds if necessary, it wasn't a likeable job, especially when they had been injured. After a few seasons I discontinued going because I felt that I was not part of the team, also I thought that many of the birds didn't have much chance.

Some birds were badly mauled by dogs. This of course was inevitable. Others where shot at close range by mistake or by the eagerness of the shooter trying to claim the bird. At the end of the Shooting Day a brace of pheasants was given to every member of the party and the meal we were given was fit for a king.

I suppose one could say that because I didn't kill the birds I ought not to eat them but I did and I enjoyed them.

HAVE YOU EVER

A sound of guns in distant woods,
    The last days of shoot had just begun
Five thousand birds were reared for the season,
    Bred to be shot, no other reason.
Have you ever heard the beaters shout,
    As they poke and prod their sticks about
And thrash the bramble bare,
    To drive out what maybe hiding there,
Have you ever seen a bird put up, and heard
    The shots ring out from all around.
"A hit", they shout, as it slumps within itself
    Then plummets to the ground.
Have you ever thought, no life, no more,
    It's dead, finished.
Never again to walk, to run, to roost, or fly.
    Have you ever thought, no more, to see, to hear,
To feed, or mate.
    For all these things have gone.
        It's dead.
        It's too late.

Chapter 53

THE REVELLERS

It was always about one o'clock in the morning and usually after a National Farmers Union dance or revellers general booze up when the young Swinnertons and the local farm friends would call on us. Whenever the lights were on at our cottage the comments would be "let's pull in, Ann and Bob are still up", "come on they'll feed us". Sometimes there were as many as twelve of them, all coming in for grub.

"What do you want to eat, eggs and bacon"? "Yes and fried bread please". So I would get from the cupboard three large frying pans, one for eggs which were our own free range, another for bacon which was home cured usually in stock and ready for eating, the third for the fried bread. I was always the chef at these times and I would sometimes cook as many as thirty slices of bacon, twenty eggs, and plenty of fried bread, also sausages if we had got any.

When the bonnets were lifted up on the AGA the hot plates would be as red as the rising sun, so it didn't take long to boil a kettle or cook a meal. On these occasions when the revellers came there was no way that we could get to bed until they'd gone, and that would be about three o'clock in the morning if we were lucky.

Fortunately for us it was a good job that booze-ups didn't happen very often, or on weekdays, because we all had out work to do and Farmer Swinnerton wouldn't have been very happy with us or his lads if he'd have known what was going on. He liked them to be back at the farm ready for work the next day, especially Robert, who was cowman, and a good cowman he was too, but he and his father were always daggers drawn, they just didn't get on with each other until finally there was a disagreement between them which led to the parting of the ways.

Robert always had itchy feet and wanted to start his own Farm Relief Service. I encouraged him in every possible way before and after he had set up, I did on many occasions go with Robert to help him with his new service. Auction sales of farm machinery and equipment were to be an additional string to his bow.

About this time he bought an old-bread cart from a couple of farm ladies who had a big farmhouse and lived at Essington near Willenhall and he asked me if I would go with him to help fetch it. We always worked together so well and I was ready to oblige. I knew before I went that it would be a day's job, just like the girder we bought for my kitchen. However, we set off in the mid-morning, did the rounds at the pubs, as usual, and then eventually we arrived at the farm to pick up the bread-cart. It had been horse-drawn and was made around 1933. It had been well looked after. Originally I think it was the property of the Co-operative Society bakery department.

Robert paid the ladies. We roped the shafts of the bread cart to his Ford Anglia and off we went stopping at a couple of pubs between Essington and Gnosall where people came out to look at his prize. Some of his mates said, "where's the orse"? or "what do you want that for"?

Our last stop was a pub at Church Eaton where we had a few drinks before going on to our cottage for a meal and to offload the bread-cart, however, whilst it was parked at my gate and harnessed to the Anglia I thought that I would climb into the driver's seat just for curiosity. Immediately, Robert jumped into the Anglia and took off with me in the seat of the bread-cart holding on for grim death as we reached a top speed of about thirty miles per hour which was fast and frightening especially for a bread-cart which had never been motorised.

However, I can claim that I have ridden the fastest bread-cart in the Shropshire Country side. After a couple of weeks we helped Robert to paint it and it became quite an attraction at future County shows.

For the next few years the lock making and service work for the Midland Bank was bringing in a good living for us and we were happy with our country life. It had paid off. We were enjoying every minute of it.

It was hectic at times especially when I was being called out on jobs at a moment's notice, which was usually the case.

There were so many commitments and although our cottage was now finished and we had landscaped our front garden, we had also made part of it into an apiary and I had got eight working hives on the site.

At the back of the cottage we had our pigsty and hen and duck pens. Opposite our cottage we had a piece of land which we rented from a local farmer for the large sum of five pounds a year, plus a couple of jars of honey. There was about one acre, on which we kept a flock of geese, a horse for our daughter Carol, three goats and two kids. One of our goats was an Anglo-Nubian, another a British Saanan and a Toggenburg. The milk from the Nubians was very rich and it had a nutty taste, which I liked very much. It was said that the milk from the Nubian and Saanan goats was comparable to Jersey and Friesian cows. Ann used to make cheese from our Nubian Goats milk, and it was known as brown whey.

So with goats and bees it was milk and honey blessed for us. On average our harvest would yield between about one and a half and two hundred weights of pure honey per season. There was no profit in small time beekeeping but it was for ourselves and we could sell our surplus stock to help pay towards the buying of new comb foundation, frames, jars and labels.

I had become quite proficient as a beekeeper; I had learned the art of queen rearing, manipulation, and swarm control. I colour-coded my queens which I found to be an advantage. I could find them more easily, and I would know which hive they had come from in the event of issuing a swarm. It was always difficult for me to make an inspection of my hives when looking for queen cells, especially when taking into consideration my emergency call outs for the bank. Of course, there was always a chance of inclement weather which was not good when opening up bee hives.

I usually tried to inspect my hives for queen cells every nine days, but sometimes I had to fall back onto the tenth day. This didn't always work out but I tried to be as efficient and precise as I possibly could.

There were times when I would come home from jobs only to find that a hive had issued a swarm and they usually would form a cluster somewhere in the garden or in the hedgerows not far away, where I would sort them out and deal with them effectively.

In the early part of spring it was great to watch the bees begin their new season, I used to sit in the wheelbarrow and watch the workers coming into the hives with their pollen baskets full.

One could almost be certain that provided that they had been wintered down properly then the queens would survive that period and would be in lay at this time.

Sometimes when a queen was getting old and her egg-laying life was in decline I would have to kill her, I would find it very difficult to do this, but it had to be done in order to maintain a strong colony for the future.

During the years of my beekeeping Ann, and our daughter Carol would take up the responsibility of extracting, ripening, and bottling our honey.

Beekeeping in those days was very different from today. For instance, farmers used to set the clover and there was no oil seed rape. In my early days there was very little spraying of insecticides and if the farmers did spray then we were notified in time to be able to close the hives on the late evening before spraying began the next day.

Chapter 54

THE ISLE OF MAN

I have visited the Isle of Man on numerous occasions and I have always travelled by boat from Liverpool to Douglas. It was a cheap crossing and very pleasant, although a lengthy one when compared with air travel.

I would park my car somewhere in the dock area of Liverpool. The crossing time was about four to five hours, and the work involved there would usually take me about three hours to complete. From then onwards I had plenty of free time on my hands which was an enjoyable break before boarding the boat at about midnight for my return trip to Liverpool. Arriving there at about 5 o'clock in the morning I would pick up my car and drive home in time for breakfast.

After making a few journeys there I began to find that this method of travelling was tiring and also that the novelty was beginning to wear off. I was again out all hours so I decided to increase the prices and try the quicker more expensive way which was to travel by air from Speke Airport to Ronaldsway and across to Douglas by Taxi.

I only had hand-luggage, which was normally no problem to me. My bag was fitted with a zip and I had a small padlock fastened to it. I carried in my hand luggage all sorts of tools and devices which were vital for my work, files, springs, key blanks, locks, picking tools, and also a small pistol drilling machine. In fact the whole bagful of my tools created a suspicion with the Customs officer there and I was called to one side for questioning.

"Will you please step this way sir", he said, "yes", I replied, and I followed him into a small room. "Is there a problem"? I asked, "no sir", he replied, "what have you got in your bag"? "Tools for my trade", I said, "I am a locksmith". "Will you please open your bag sir"? "Yes", I said, and as I opened it a bag of sandwiches dropped on the floor. Making my apology I picked them up and bundled them into the paper as quickly as I could, and I put them on the side of the counter. At the time I was in a complete fluster, because my plane was due for take off at any moment. I was soon cleared by Customs, who helped me to repack my bag and was to be on my was until I heard a voice call "You have left these sir", they were my sandwiches. I didn't really want them. I thanked him as I pushed them into my bag, 'but' as I closed it they got caught up in the zip, and there was cheese and tomato all over the floor. It caused quite a laugh with the Customs man.

I think that they thought that I was carrying a gun and a load of ammo in my bag. However, I finally arrived at the bank in Douglas. After I had finished my job I couldn't have faced a bag of mashed up sandwiches, so I gave the seagulls a good meal and went for one myself.

I remember this job very well because of an announcement made over the speakers on the 16th of August 1977: Elvis Presley is dead. Passengers were devastated; it happened to be my last visit to the island.

Chapter 55

BLACKPOOL HIGHLIGHTS

Arriving at the bank in Blackpool was quite a hectic night for me in more ways than one. People were in high spirits, and the Illuminations were in full swing. I arrived in the evening about nine thirty and I had a lot of work to do because the cleaner had lost a key to the front door rackbolt lock.

I was lucky to be able to park my car and the police looked after me well; I could stay as long as was necessary. It was very noisy about town, and the front was very busy with good- timers, enjoying themselves on every corner including the one where I was going to work.

I introduced myself to the messenger who was waiting for me. He wasn't happy because I was late due to the heavy traffic coming into Blackpool.

It looked as it I might have difficulties working on the front doors especially with the crowds about.

All locking securities had to be moved from the door for much of the time. People were pretty good and made the usual corny but harmless jokes about the bank being open at that time of night, it was unheard of in those days, comments like "how's it looking for a few bob", or "it this is a break in I'll give you a hand".

It was very late by the time I had finished there. I tried all the new keys, locked the door and went through the Banking Hall into the back of the branch. The messenger made a cup of tea after which I was ready to leave, when we heard a continual rattling on the door. "What the bloody hell is that"? said the messenger as he hurriedly made his way towards the front door. "I don't know", I said, "the doors are locked".

He looked through the viewfinder and he couldn't see a thing but the door was still rattling well. He picked up a chair, stood on it, got down again and said, "get up there and have a bloody look at that", there was a couple having it off against the bank door. No wonder we couldn't see through the viewfinder they were covering it up. "We'll soon shift 'em", I said, "open the door and they'll fall in". "No" he replied, "give them a break let them have a bit of time while we go and have another drink." I thought he's a good sport and didn't want to spoil their fun, or was he thinking of his overtime.

They were "having it off" and I wanted to get off. I had a long way to go.

Chapter 56

THE JACKDAW STORY

In early April of 1982 two friends of ours, Maurice and Vivian came to see us and asked it we would be able to help in trying to save the lives of two Jackdaw chicks.

They were not fully fledged and they had been taken from their nest in a chimney at the rectory in Norbury, which was two miles from our Cottage. The parent birds had forsaken them because builders had been working on the chimney there.

Maurice and Vivian had been feeding the birds for a couple of days on white bread and water. We didn't hesitate in taking them on and we insisted that they fetched them as soon as possible, because feeding birds on white bread was just an existence and they would not survive for long.

Our friends went to fetch the two chicks from their home in Newport and within the hour we had got them.

We had to work very quickly because they were deteriorating rapidly. As soon as Maurice and Viv had gone we jumped into the car and went the Angler's shop in Stafford where we bought mealworms and fresh maggots and hurried back to our chicks. We kept them in a cardboard shoebox with some sticks and straw and they were comfortable.

Feeding times were very regular for at least the first two weeks. We used to go to Stafford every day to get fresh worms and maggots. Ann looked after the feeding and she had to devote her time to it. They always wanted more until their tumms were full. Ann didn't like picking up the grubbs with her fingers and dropping them way down into the bottom of their ever-open beaks, so I bought her a pair of tweezers, which worked admirably. They were now having plenty of food and a special "fill-up" before dark. We kept them in the spare bedroom at first so that we could hear them call out at dawn for their grub.

They were coming on well and as they got a little older we moved them into the garage with bales of straw around.

We still fed them on their meaty diet with sometimes very small pieces of cheese and also Weetabix, which was to become part of the adult diet for the one bird.

Up to now we were very pleased with our efforts and it was time for us to teach them to fly. To do this Ann and I used to have the birds on our outstretched arms, we ran around the garden dipping and raising them slightly, this helped to strengthen their wings. They were progressing brilliantly until one day when our dog Bessie got one on them, I was very quick to make her drop It, but unfortunately it suffered a broken wing, I took it to my vet at Newport.

He supported the wing with a tight fitting sleeve but it was not successful and the bird had to remain in captivity for the rest of its life, from then onwards we named it Jackie. The other one we called Jack.

I feel sure that had "super-glue" been around at that time things might have been different for Jackie.

However, I built an elaborate bird house for them both and Jackie lived there for twelve years. During this time we had taught him to talk. He lived a life of luxury until he died in 1992, and was buried in our cottage garden along with many other animal pets.

We put a ladder up to the birdhouse and we also put another one against a damson tree in the garden so that both birds could use them. Sometimes Jackie would stay in the damson tree all day whilst Jack would fly around the garden and fields, but always come back to the tree or the birdhouse.

Ann with one of the Jackdaws.

Jack made excellent progress and by the beginning of July he was ready for the wild, but he didn't seem to want to go and he stayed with us until the March of the next year.

As they grew older we found many characteristics and affections in them which were unbelievably unique. For instance, whenever Ann used to go on her bike to the village of Gnosall or on foot which was usually about two or three times a week, Jack would perch on her shoulders and go along so far with her just for the ride, after which he would fly back home.

There were also times when Ann would walk and not take the bike with her, on these occasions, I noticed that Jack would fly onto the television aerial and perch there. The bike was not connected with the situation in any way, and it was obvious that he was "on watch" awaiting Ann's return journey home. It was so interesting and I was determined that I was going to find out a little bit more about it. I sat down for a long time outside, continually looking up and watching his mannerisms until eventually his tail began to flick and he became excited.

I had my binoculars with me and I hastily made my way to the upstairs bedroom window where I could scan the lane, I soon found Ann. She was walking past the farmhouse gate about half a mile away from our cottage, and then I knew that Jack had found her too. Although I couldn't see him land on Ann's shoulder, he was there.

How he knew who it was walking the lane from a half mile away I shall never know.

Another regular habit of his was to fly in from the fields around the cottage, down onto the kitchen window ledge, tap on the window for a digestive biscuit and a piece of cheese, then he was off again. During the following March he left home for some months when it was obvious that he had found a mate.

He always returned and stayed a while during the very cold of winter. This went on for a number of years until one day when he left and didn't return.

We thought that he must have died, or been shot and that was the end of our Jack until one very cold winter's day. After about two years absence we were having our breakfast in the kitchen when there was a tapping on the frozen window pane. It was Jack! We couldn't believe it, he had come back to see us and for his biscuit and cheese. He was very hungry and he stayed with us until the early spring when he left. We never saw him again. We had devoted a lot of our time to Jackie and although he couldn't fly very well, he could get about the garden where he loved to perch in his damson tree. He was a very happy bird. He could talk, and unknown to him, the way

I answered him created an understandable conversation, which only Ann and I were familiar with. We enjoyed immensely.

When he called for water it may have been a coincidence but I always noticed that his bowl was usually empty. On the other hand he could have associated the sound of the word water after hearing it so many times and also seeing us fill his bowl.

He was a great bird to have around; I remember when one day we almost lost him. He was out in the garden when a buzzard attacked him Ann and I ran into the garden making an awful noise to frighten it off. There were many times when Jackie would go on Ann's shoulder across the field or down the lane, he would never leave her. Even when she was digging the garden he would hang around looking for grubs.

As I have previously said Jackie died naturally during 1992, and that was the end of the Jackdaw story.

Jackie's favourite words: -
- Jack
- Jackie
- Alright
- Later
- Water
- Yes
- Good morning
- What are you doing?
- Wait a minute
- Do it now
- OK

Chapter 57

PORTHMADOG

I had visited the bank at Porthmadog many times in the past and I was familiar with the front door locking system there.

It was a rackbolt lock manufactured by a company known as Adrian Stokes. The lock was quite complicated and troublesome, and there was room for design improvements.

In general it was not accepted by the bank. Although they had used quite a few of these locks, they were scattered in various parts of the country.

Some were in operation in the area of Cheshire and North Wales; others were in the North Country and the Midlands. They were quite difficult locks to repair and most locksmiths would not get involved with them.

Worralls and I knew from experience about the pitfalls relating to the repairing and fitting of the locks. Outside London I was the only locksmith coping with this situation. I had made plenty of spare parts and I always carried them with me although I could always make them on site if necessary.

On my arrival there, I was immediately recognised by the branch accountant whom I had met on previous visits. We often had a chat for a few minutes before I started work. Although he knew me I would always have my ID ready to present to him before getting the go ahead, because it was bank procedure. I always removed the complete locking system from the door, rods, cover plates, bearings and the lock itself, and I laid the parts neatly on the floor of the bank beside me. I had my tools on the step and it was obvious to anyone that I had the authority to be there. Suddenly whilst I was working the alarm went off, it wasn't anything do with me so I carried on working, and then after a few minutes I heard the sound of a Police siren. I saw a Police van coming quickly and it stopped outside the bank.

Two officers, a Sergeant and a Constable, jumped out of the van. The Sergeant came to where I was working. He was very abrupt in his mannerism, and said, "What do you think you are doing"? I was dumbfounded and I couldn't answer him for quite a few seconds, and then he snapped and said, "I asked you a question, what are you doing"? It was obvious to anyone with common sense that I was working on the lock; also customers were coming in and out of the branch. I was irritated by his attitude and snapped "what does it, look like? Can't you see I'm repairing the lock". "There is no need to be clever with me" he replied, "Where is your identification"? "In my pocket, where's yours"? "Let me see it", he said. "No", I replied, "you go and ask the accountant, he is satisfied with my presence here". Then without any hesitation he caught hold of my arm and tried to lead me quite forcefully to the enquiries counter. His attitude was that of an arrogant bully, and from then onwards he had met his match. I dug my elbow into his side, and said. "Take your bloody hands off me". Police or no Police my blood was boiling and I wasn't having anymore. I called the accountant, "will you tell this man who I am", "Yes", he said. Whilst they were talking I took the opportunity of saying to the sergeant "I am going to get on with my work, I've got no time to waste with you". To which he replied. "I'll come and see you in a few minutes". "OK" I said, "and I'll be bloody well waiting for you". I thought perhaps he was going to arrest me. However, if he were, then it would have to have been after I had finished the job, because they couldn't have locked up the bank, and they may have had some real bank robbers on their hands.

The Sergeant came to see me at the door where I was working. He wasn't quite so abrupt, although he did try to caution me, but I wasn't having any of that. I told him to say what he had to and get out of my way.

About twelve months later I was recalled to the branch and I met the accountant again. We had a chat and he told me that the Sergeant had been given a reprimand, and also he had been moved on so it appeared that there were other problems there.

Chapter 58

A MEMORY FOR A VOICE

Many of the unusual things which happened to me always seemed to be when I was out on jobs. On this occasion I was to go to work at the Midland Bank at Hawarden in Flintshire.

I was to carry out some repair work and supply new keys to the front door lock. After working at the Branch for about two and a half hours I called for the attention of one of the Bank Staff. I've finished the job I said. Would you please try the keys for me, and sign my notes. "Yes certainly", she said, "I suppose you want to get off home, have you got far to go"? "No", I replied, "only to Newport in Shropshire". As we talked I heard someone call out "I know that voice", and again "I know that voice". I looked across and I saw a man walking into the Banking Hall from the direction of the manager's office "I'm right" he said, "I knew it was you, I could tell your voice" I looked puzzled and yet I knew his face. I had seen him somewhere but where. "I've been to lots of banks," I said. "No, no before then, thirty-five years ago, you were in the Army, 1944" "Yes, you are right" Royal Engineers, Ruabon, North Wales, Wynnstay Hall, and you come from Wolves". "Yes, right again, I said. "And your name is Sidbotham, Sapper RE. Known as Sid, to all your mates - well" he said "meet Taffy Hughes again after all this time; you and I were on guard together quite a few times at Ruabon. You haven't changed a bit" he said, "I could have picked you out in a thirty thousand crowd at a football match"!

It all came back to me in that short space of time after our chat over another cup of tea. Then on my way home I was able to think more clearly and when I looked at some old pictures I found a couple with Taffy Hughes on them. It had been a great day for us and we kept in touch. We did meet a couple of times when he went to manage another branch of the bank at Llangfni on Anglesey where he lived and retired. I always wondered how anybody could recognise another person by memory of voice alone. With a memory like that Taffy Hughes must have been a very good Bank Manager.

Chapter 59

THE CHILDRENS COUNTRY HOLIDAY FUND

Our daughter Carol had just had her twenty second birthday, and boys had been coming in and out of her life quite frequently until she met Pete. They went out together for twelve months before they married in 1980.

Just after their wedding we decided that we would apply to the Childrens Country Holiday Fund and offer our home for the school holidays to some of the deprived children living in London.

It wasn't long before a representative came to see us at our cottage. We were soon accepted, and were ready to become a host family for the children. After a few weeks we were on our way to Wolverhampton Station to meet our first guest. Before we left our Cottage, Ann and I went round the hedgerows making two or three nests in which we put eggs from the henhouse so that with a little help the kids would find them. This was to help them to settle in, get to know us and find their way around. From then onwards it was plain sailing for the next ten years. During those years it was most enjoyable for us to watch kids having such a great time; kids that had never seen a farm or farm animals before.

They loved helping the farmer in the fields at harvest time. Our stock was also good fun for them. They felt very important when milking the goats, collecting the hens' eggs, and of course riding the pony. We were very lucky living in such a wonderful place, and were very happy to share it and devote our time to these kids from Stepney.

Each year we would take many photographs of the children, put them into an album, and post them on to them after they had returned home, so that each child had their own album for a keepsake, memories of their country holidays with us. Although these kids came from poor homes we never ever had a spark of trouble from them in all of the ten years that they came to stay with us.

On the Christmas Day of 1984 our only grandchild was born, Victoria. I call her Bunji, when she was three years old she used to come to the cottage and play with the girls from Stepney.

Unfortunately for me during March 1985 I was diagnosed as having a cancer of the prostate, which meant that I would have to slow down considerably at least for a time anyway, and although we still carried on with the Children's Country Holiday Fund until 1990. It was then that we decided that the next child would be our last one. So we thought that we would make our exit by holding a charity auction.

The preparation was quite hectic. We went from farm to farm begging anything from half-pigs to sheep and beer from the pubs. Shops and businesses in and around Newport were most generous. On the evening of Friday the sixteenth of October 1990 we held our auction at the Village Hall in the small Village of Cheswardine in Shropshire.

Our auctioneer and great friend was Bob Swinnerton, he was a self-taught auctioneer and could sell arrows to the Indians!

It was a grand evening: we had four hundred and fifty lots to sell. Everything from a three-piece suite to a tin of fish and we sold the lot.

It was hard work scrounging, and especially hard work for Bob. He worked for five hours non stop and the total raised was over six thousand pounds, which was divided equally between the Children's Country Holiday Fund and The Shropshire Deaf Society.

After the presentation, which took place at the Red Lion Pub at Wistanwick in Shropshire, we decided to retire from the CCHF to which we had given our full support over the years and we had enjoyed every minute of it.

Chapter 60

THE FUNERAL

On Saturday the 11th August, 1984 our good friend and neighbour, Frank Swinnerton died leaving his wife, five sons, and a daughter. Like most country funerals in the farming community there were usually many followers, and it was accepted that this would be so at Frank's funeral which was to take place on Saturday the sixteenth day of August 1984.

There were over four hundred mourners present on that day. Friends from outlying villages and farms, relatives from different parts of the country, all had come to pay their respects to a farmer whose funeral was to become a legend. For reasons I shall not go into, it was an embarrassment and a most upsetting occasion for the family and their friends. In summary, it gained the attention of the media and was referred to as the Cheswardine Hullabaloo.

Going back to March 1985. I was like most cancer sufferers, under stress awaiting results from further tests to find out whether I had got any more complications. At this time I needed something to take my mind off these uncertainties, which were hanging around me.

I had my lock work, but I was looking for something else, another interest, which may help me over this bad patch. I found it when looking at a programme on the television, how to make a wild life pool, by Chris Bains. A wild life pool that looks good, just what I needed to do. Plenty of digging and something to show for it. I didn't hesitate in sending for the details and although I had the bees to look after, they didn't require any attention for a few weeks.

As soon as I received the plan, it was both hands on the spade. The area was going to be quite large with a stone part for frogs to climb in and out, a marsh and many other features were illustrated on the plan.

Around the pool I planted meadow sweet, marsh marigolds, bulrushes, poppy's, harebells, and many other wild flowers.

I remember one day very well when I was working in the garden. My doctor telephoned to tell me the good news that I had no secondary problems. Although I was going to have to attend hospital for six weeks radio therapy treatment I considered myself very lucky. During and after the treatment I was able to finish my pool. It had matured after a couple of years, and it looked great until I got up one morning only to find that it had been invaded by moles, big moles, super moles. They played havoc in the vegetable patch and around the pool edge.

Like most country gardens we did have moles, but we always managed to keep them under control with the use of traps. It was hard going this time: the traps weren't working. I remember catching one just as it tried to burrow its way into the grass I got hold of its stubby tail and pulled it out. I took it down to the brook to let it go. It bit me, and it continued to bite into the palm of my hand, I was quite worried, I squeezed it quite hard until it let go. Needless to say I never picked up another mole unless it was dead. After a few days I declared war on all moles in my garden and I reverted back to the 'twelve-bore' which I had used in the past. It was sure quick, humane, and I never missed.

My method was to watch for the movement of soil on any mole hill in the garden, then I would fetch the gun. By the time I returned with the gun the movement of soil had probably stopped but not for long and I had a lot of patience. I would stand there sometimes for half an hour or more with a loaded gun, waiting for movement. The battle of the moles went on:

TEAR AWAY

I am the mole, who makes the hole deep in the grassy green,
I work so hard, the mounds I make?
The best you've ever seen,
When I've had just what I want, I move myself around.
I'll have a go. The vegy patch. I'll tunnel underground,
The seeds are sown so straight just here,
Plants are looking good.
I'll bulldoze my way through the lot,
I will. I know I could.
Four mounds I've made this very day, and tunnelled round the pool,
Two traps I saw, I dodged them both.
You know. I am not a fool.
So sure I was until the end, whilst working my last mound
A blast oh such a blast there was,
It blew me from the ground.
Perhaps if I'd been moderate, I'd be alive this day
Had I thought he'd use a gun,
I would have moved away.
Who ever heard of a moderate mole anyway?

Chapter 61

SECRETS IN THE SAFE

I received a telephone call from the principal of a college in Shropshire who asked me if I would open a safe for them, and would I visit the premises beforehand to give them some idea of the costs involved.

I was quite happy to do this because it was only about twelve miles from where I lived.

It was a very good quality safe and was manufactured by a well-known company in Leipzig in East Germany probably made in 1935 or 1937.

I asked about the possibility of finding the keys, but I was assured that there were no keys to be found. I established that to the best of their knowledge the safe was last opened during the war some forty years back, and that the person in charge of the safe at that time must have been in possession of the keys. I was told that he had been dismissed from the college for reasons best known to the college authorities, I was also told that it was thought that because the keys had not been found that he had locked the safe and disposed of the keys or taken them with him for which there seemed to be no apparent reason. It was a strange situation and became more interesting to a couple of members of staff who obviously were inquisitive regarding the contents of the safe. However all was not revealed until a few days later when, I returned and opened the safe and the contents had been examined. The lock was very robust and well protected, it was difficult to open and took about two hours. On opening the door we found that the safe contained many papers, some of which looked official and were written in German, also there was a stamped document which I thought to be that of the Third Reich.

It appeared that we may have opened the door to some secrets forty years locked away.

However when I went back to refit the lock and make good the repairs I was told that the papers had gone to London for further investigation. It appeared that he was spying for the Nazis, which would certainly provide an answer to the missing keys.

Chapter 62

BURNING OF THE STRAW

After the grain harvest there came the baling of the straw. Wheat straw was used for bedding, spring barley straw for animal feed.

The field at the back of our cottage was quite a large field about twenty-four acres in all.

Harry, who farmed the land there, would bale the straw and keep it for his stock, and any that was surplus to his requirements he would sell off to the highest bidder.

During the early 1980s nobody wanted the straw. It cost too much to bale it, so it was put up for sale on the field and then if it didn't sell it was burned off with the stubble.

Firstly the firing took place around the edges of the field in order to keep control of the situation, because the heat from the flames was so intense.

After the burning of the straw there was a black stubby dusty ash all over the field. When it rained it would become a soggy black mass. It was said that the straw was burned because no one wanted it, and that the potash was most valuable. Firing the straw was also a way of killing bacteria in the ground.

All these explanations were probably true but I think that if Harry, our farmer friend could have sold it, he would have done so, which would have been for the best anyway. However whilst burning the straw at the edges surrounding the field the wildlife became trapped, and was roasted alive, it was a most horrible sight. Ann and I had walked the field directly after the fires had gone out, hedgehogs, rabbits, frogs and many other animal remains were there. I was prompted by this to write a small poem.

THE BURNING BLACK

Run creatures run, one and all
Run spider large, run spider small.
Run fast, run fast, no looking back.
There's death within the burning black.
Where creatures lived
Where creatures born,
amid the ripened golden corn.
How black the smoke, now burns the straw.

Left from the harvest yield.
Life has gone within the flames
That sweep across the field.
Now begins the farmers toil
He ploughs and works the blackened soil.
Covered by silver threads
Spiders make their lacy webs
I see again the plough this morn
Where soon shall grow the winter corn.
Life is born anew once more.
Since last, the burning of the straw.

Chapter 63

BLOOD ON THE SNOW

During the summer of 1986 one of our CCHF children asked if she could come to spend Christmas with us. Ann and I talked it over and we agreed to do this. It was to be our very last time, a one off because we had made a promise. We made the necessary arrangements through the authoritative body in London.

Just before Christmas we met our guest at Telford Railway Station. We returned home to the cottage where she stayed with us for almost three weeks.

Life was very different for her in the country at Christmas time than it had been during the summer holidays over the past years. We thought that she may have been homesick and missed her sisters at this time of the year. However, this was not so.

We did our best to make it an enjoyable holiday for her. We prayed for snow and we got loads of it. It turned out to be a super Christmas for us all, but not too good for the local farmers and publicans. They were without power and lighting for a few days and the milking had to be done by hand.

Our local pub owner Jim kept the Junction Inn at Norbury and knowing that we had generators and paraffin lamps he was soon on the telephone "Bob, can I borrow some lamps off you", "yes Jim, you can have four, is that OK"? he was soon to come round on the tractor to collect.

Needless to say for us our generators were working in full swing and we were the envy of all around. Our guest was enjoying Christmas to the full She was very fond of all animals and looked after our stock like she had previously done in her summer-time visits until one morning when she came hurrying back shouting "Uncle Bob, Uncle Bob, come quick there's blood on the snow". There was blood on the snow all right, our hens had been attacked, Reynard had struck again. We hurried through the wicket and across the field to look at the damage done. It was a massacre, dead hens everywhere, heads bitten off and feathers littered the snow.

We always thought of these happenings as being part of country life, but I feel sure that it would have been much more acceptable if Reynard had killed to satisfy his own appetite. Ann was very anti-hunt, and although she knew that there was to be a meet in the area during the next week she was very upset about the attack on the hens but she still hoped that Reynard would escape from the hunt.

BADGERS COVERT

A caller had left his vicious mark,
'Reynard', the cunning of the dark,
Looked like a battle had raged that night,
They were roosting hens; they'd make no fight
Chickens but a few hours ago,
Now scattered feathers, heads, littered the bloodstained snow.
Tracks from the covert, to the woods we had seen,
They were the tracks where Reynard had been,
We followed some way for what it was worth,
No sight of the Fox, he's perhaps gone to earth.
A massacre at Badger's covert, Reynard had struck again,
His killing, a gruesome site,
The morning after the dead of night.
There's talk of the hounds coming this way,
And the huntsmen of course, its their sporting day.
'Reynard' the cunning, he'll hide or lie low,
Unlike the chickens, they had nowhere to go.
If he stays were he lives. His chances are good,
For the cover is dense in the heart of the wood.

PART III

LOOKING FORWARD
TO RETIREMENT

Chapter 64

HOLES IN THE WALL

Willenhall, the heart of the English lock trade, was known to some people in and around the Black Country as "Humpshire". It was said that the old lockies of years ago became either round shouldered or they developed humps on their backs. This condition was thought to be caused through standing for long hours over many years working at the bench and vice.

After six long days many of them would start work again at home in the late evening either in their sheds or washhouses for extra cash. Beer money was a means towards their social existence; it was said to be essential if only to swill down the brass dust inhaled during their daily labours.

In the mid 1930s, as a young boy my Mother would take me to see my Dad, who worked in a little old sweatshop in a back street in Wolverhampton. He had worked there since about 1925 and his father had worked with him for a few years before his retirement.

Dad was then about 38 years old and he was the youngest man working there. The firm was known as Ely Ling & Son of Lewis Street, Wolverhampton. There were four other employees, three of them were in their mid to late sixties: Big Ron Taft, Little Frankie Flynn, old Syd, and old George Broom. Of the four, as far as I can recall only old Syd had a "hump!"

I remember seeing these men at work covered in brass swarf. George Broom kept in touch with Dad. He was one of the old faithfuls who came to see us at our factory during 1948, where he would chat about past experiences. Dad used to talk about Willenhall being called "Humpshire" and about a pub I think it was called The Bulls Head, near the town centre. Supposedly, there were holes built in the wall where the old lockies could rest their humps, tell their tales and enjoy a pint in comfort. Unfortunately I have found no evidence to support this, but the name "Humpshire" persisted for many years.

LOCKSMITHS TOIL

Me Grandad were a Locksmith,
Me Dad 'e were one too.
I follered in both on 'ums footsteps
T' learn the trade they noo.

They used t' tell their tales o' many 'ears agoo,
I 'erd a one – remember it well – a pub in Willinall
Where Lockies used t' sup, and rest their 'umps
In 'oles med in the wall.

There wuz a bloke node as Franky Flynn,
'e always used t' say,
'ar con mek um Gaffa'
Ah yo con mek um it they'll tek um, but will they bloody pay.
Every Mundy mornin' eed goo up fer a sub
P'raps five bob or ten t' tek 'im t the pub.

Sid wuz another mon oo always luv'd is ale,
'e said to 'er one day when 'e cum 'um frum wirk,
"tay fit fer a dog t' be out t'night, I aye 'arf glad 'arm 'ere,
Get the jug, goo up the pub 'n fetch a quart a beer.
In the end poor ode Sid, 'e went down with the blues.
Never did recover t' get back on the booze.

I remember when I started, off t'wirk I'd goo
Me samwidge wrapped in pairper, me airprun white as snow.
I couldn't use the 'ubbin, the ommer, or the saw,
'acksaw blairds, 'n files, I'd bost 'em be the score.

Oh t' be a Lockie wirkin fer is Dad,
It wore the brightest future that a fella cud av 'ad
I'd screw me kay oss in the vice, and then I'd round me kays,
It tuk me 'ears t' learn the skills in them ode fashioned days.
We slotted lavers all be hand, shairped the bellies too,
Screwed the caps down on the locks and 'oped they' bloody goo.

Some 'ears 'ad passed, and then at last
I mastered that there trade.
It's knacks, it's skills, I know 'em now, a locksmith I ave made.

Forty 'ears a Lockie, taught the ode ode way
Never med a fortune, never 'ad much pay.
Cumin't' the end now, hand med works nigh gone,
T' me it doe much matter
Cuz in five 'ears I'll be dun.

Chapter 65

HARD TIMES AHEAD

The next few years were very hard going, jobs on the road were plentiful, but as usual there was a shortage of service engineers especially in the banking world.

At the time there was no way that I could cope with the amount of work I had on hand, I was not fit enough, my working days were numbered. I was lucky to be here, and I knew it.

I carried on under very arduous conditions until 1991. Jobs became increasingly more difficult for me to do. The side-effects from radio therapy treatment had caused the tissues to scar in and around the prostate area, making driving extremely uncomfortable especially over the many miles which I had to travel, this together with the hours that I had to work on site, became an aggravation which I find very difficult to describe.

I had commitments regarding the bank. However some of the lock making side of the business I was able to hand off to my cousin who had a small lock factory in Willenhall. This was good and it provided me with a fair amount of profit.

The other work I had to cope with myself and be on call twenty-four hours a day as I had done over the past years.

I had other responsibilities too, such as our cottage and garden, which Ann and I had tended and maintained for years. It was very expensive to upkeep but it was our baby, and it had been a good investment for our future.

There was also my private pension along with our other insurances, which had to be paid for. It was on the fifth of July 1991 that I reached the grand old age of sixty five, and although I had used a lot of my savings whilst I had been ill, I was now able to reap the benefit of some of my insurance investments.

It had been a struggle to keep them going, and now I was on the receiving end.

Fortunately the time was right for me, just as the bank were moving into another era regarding their security systems, I was able to scale down some of the work that I had with them, and at the same time increase my prices which gave me a much higher rate of profit, this enabled Ann and me to have some holidays abroad, also we could continue to maintain our cottage and garden, keeping it in the condition which we had been used to doing.

It was in June of 1995 that I decided to pack up lock making altogether, enough was enough, so I discontinued and retired. I will always remember my last big job; it was a journey to the Midland Bank at Redruth in Cornwall. Twenty hours none stop, too much for any man whatever the age. It was a good job that I had made my mind up to finish work, because I had a feeling that I was having another flare up in the area of the prostate.

I always went to see my Doctor whenever I thought that there was a significant change of pattern. On this occasion I was right, there was something that had to be checked. The cancer may have reared its ugly head, and could be on the move once again.

I was quickly referred back to Stafford General Hospital and soon admitted for further investigation.

During 1995 it was decided by the consultant urologist there, Mr. Frank Murphy that I was to have another type of treatment. It was to be an injection into the gut wall, a hormone called Zoladex and was to be administered at the Doctor's Surgery every twenty eight days, for life, of course I was over the moon. To think that I had been given yet another chance, so I jumped at the opportunity.

After being weaned onto it for a couple of weeks with tablets, I was ready for my first injection, which took place on the seventh of June 1995. For the next six years it was highly successful, and whilst I knew that there was no cure it was good to know that something can be done. The side effects from the radiotherapy were always there plus other side effects from the Zoladex injection.

There were also times when I couldn't sit down, and I would have to sit on a rubber ring to ease off the pressure to try and make sitting more comfortable. It is difficult to gain comfort for any length of time without fidgeting about. Although I have problems I am still knocking on well, thanks both to Stafford General Hospital and to our Doctors at Wharf Road Surgery in the small village of Gnosall in Staffordshire with whom I am able to play an active part in fund-raising for the patients of our Surgery. Seventeen years with prostate cancer, still here, can't be bad.

A golden rule for me, if in doubt consult my GP.

Do not delay, ask your GP for a P.S.A.

Chapter 66

OUR VILLAGE PRACTICE

I have been a patient at our village practice for thirty seven years. All of the G.P.'s and staff there are, and have been very highly thought of by the patients of Gnosall Surgery and its surrounding villages, of which there are about seven thousand inhabitants.

Many patients including myself come away from the Surgery, sometimes without medication, feeling much better than when they went in. Obviously a boost from the friendly Doctor had a good effect on the well being of the patient; "well" certainly it did on me.

However getting back on track again, during 1996 there became a problem in the village of Gnosall that could have changed the lives of the patients of Gnosall Surgery.

It involved drastic changes in the dispensing of medicines and drugs. A chemist was coming to our village. There had never been a dispensary in Gnosall other than at the Surgery. Gnosall Practice has always had very high standards. Profits from the dispensary were ploughed back into the system to help to contribute towards the funding of an extra Doctor for the patients needs, but this may soon come to a swift end. A chemist was coming to set up shop somewhere in the village.

Now had this happened it would have meant that our Surgery would have had to close its dispensary to all patients living within a one mile radius from where the chemists decided to site their shop. Immediately there was to be a meeting, it was called at the Village Memorial Hall. People came from Gnosall and its surrounding Villages to make their protest, the hall was packed and the people were not happy.

There was no way that patients were going to concede to a chemist's shop here. They advocated freedom of choice, and that was not in the direction of a Pharmacist coming to Gnosall, and so the protest was on. The situation was put to the people during that evening by Barry Williams a spokesman for the Doctors. However when he informed them of the possibility of a loss of a Doctor the meeting became more aggravated, the heckling began and the support was almost one hundred percent.

Dr. Greaves was leading the Practice and made his way to the stage. He quickly put forward the future of the Practice should a Pharmacist come to Gnosall. He went on to say that the profits made from the chemists shop would be taken out of the village instead of being ploughed back into patient care as it is now, and has been in the past.

After Dr. Greaves had returned to the back of the hall Barry Williams put on screen the facts about this situation, and called for questions to clarify different points of view. He also proposed that a "support group" be formed to put pressure on the Government to try to change the existing law of the restriction of the one mile radius, thus giving our patients a freedom of choice as to where they would wish to shop should a chemist come to Gnosall.

A change in the law would mean that our Doctors could retain the rights to supply medicines from our surgery.

Doctor Mulligan sat on the left of the stage hoping to form a group which he did that evening.

I joined without hesitation. I knew from past experience over the years that it was vital to our patients that the present situation be kept whatever the cost. So I jumped in at the deep end to join the fight.

A show of hands was called for and there were plenty.

Dr. Mulligan took all names and our group was formed: "Gnosall Surgery Action Group", and a highly successful group it turned out to be.

At the end of the evening there were around twenty-four names put forward.

The next move was to arrange our first meeting, which was held at the surgery, the purpose of the meeting was to discuss our strategy and allot jobs to various members.

Dr. Mulligan was on the group for his professional expertise, and also he was one of the crew. Barry became spokesman working along with Roy Barratt who was in the chair. I went along with the rest of the team volunteering for anything that was going to help our cause. After a few weeks Ann decided to come along and join the group. At that time it was necessary for us to have a reasonably accurate figure on how much backing we had from the patients, and although we had good support at the village hall there were still over six and a half thousand patients out there of which most were legible to sign a partition. It was essential to present an accurate account of the situation to Government. One of the first important jobs was to tramp around Gnosall and its surrounding villages getting supportive signatures. There were about seven thousand patients in all, of which well over four thousand were eager to sign.

Cheque presented on 4th September 1996 for £4,014.50, following funds raised at local auction for Gnosall Surgery Action Group. Left to Right – Bob Swinnerton (Auctioneer and friend), Ann Sidbotham, the Author and Roy Barratt (Chairman).

Whilst we knew that a chemist was looking for premises in Gnosall we continued to put together our strategy to try to keep him out. Our group meetings were held every month; there was a lot of work to be done. Our Chairman, Roy Barratt, was doing a good job; he was also trying to raise funds for legal expenses but he had found it too much to do both jobs.

So at one of the meetings he put forward a request for a volunteer to carry on with the fundraising activities. This was a job that Ann and I had been used to doing, so I put my hand up. "Ann and I will do it, we'll take it on, but on condition that we pack it up if it gets too much for us".

We had to raise a substantial amount of money quickly to help with expenses and legal fees so Ann and I decided on an auction. "Where are we going to hold it"? she asked. "I'll go and see the School Head Master, Mr. Butlin", he was very eager to help, and we soon got sorted out.

The main School Hall was ideal for us. We booked it three months in advance of our requirement, which gave us adequate time to collect and put together the whole affair.

We went from house-to-house collecting everything from an electric typewriter to an old sink. Farmers were most generous and sympathetic to our cause, they donated half pigs, sheep, etc.

I wrote to ex customers of mine for their support; they gave wall safes, and locks. Shops in Newport were also very generous. All of the goods were stored at our cottages, and the rooms were full including the garage and out houses.

Then came the great day on Saturday the twenty fourth of August 1996. We moved all of the stuff down to the School and were ready to start selling at seven o'clock in the evening.

Our good friend Bob Swinnerton was to do the selling. By 11pm almost every item had been sold, we had raised over four thousand pounds!

It gave us great pleasure to do this and it was very rewarding especially for me, because I had been able to put something back into the system, which had been so much help to me in the past.

Chapter 67

A 10K WALK

Our group membership was steady. Like most groups it had thinned itself out and we were left with supporters very dedicated to our cause. There were now fourteen of us and we continued to raise more money for legal fees and there was talk at this time about taking our fight to Europe. Fortunately this didn't happen. Soon after the auction in August we worked out our 10K sponsored walk. The starting point was at Gnosall where we were able to walk along the old Railway embankment after which we were to take the canal towpath to Norbury "The Junction Inn", at the Shropshire Union Canal Basin.

Our return journey was a short walk through Lord Lichfield's estate to join the old railway line again at a different place and continue through the woods towards Gnosall. It was on Sunday the thirteenth of October 1996, the weather was gorgeous and we raised just over one thousand pounds.

These two successful fund raising events gave us a substantial bank account and we were ready to fight our legal battle with a bit more punch.

Chapter 68

FIGHT THE GOOD FIGHT

1996 was a busy time for the Action Group. The chemist had not yet come to our village. During the early part of that year it was decided and approved by the Doctors at our Surgery that there was to be a shop set up within the Practice.

The shop was opened in June 1996, just before our first auction, and it was called Gnosall Health Care Shop. It turned out to be quite prosperous, and was well patronized by the patients; it sold health drinks, nappies, powders, creams, and many other items including toiletries, all kinds of medicines and tablets that were permitted to be sold without prescriptions. It was run very successfully by one of our group members.

Initially, the shop was set up with a view to apply for permission to open a Pharmacy within the practice at Wharf Road, Gnosall. However, this was not to be.

As we continued to push on with our work, we were gaining support from the patients; in fact by July we as a group had obtained four thousand signatures; this was a wonderful achievement thanks to the patients of Gnosall Surgery. The petition was taken to Parliament, delivered by hand, where it was supposedly lost, mislaid, or whatever, but that's another story. Our last bit of fund raising for 1996 was on Christmas Eve when the group had decided to go Carol singing in the village. We asked a local builder if he would lend himself and his lorry and drive us around Gnosall.

We decorated his truck with bales of straw, holly, and balloons. One of our group members, 'Don', volunteered to dress up as Father Christmas, and he looked every bit the part as he sat on the back of the lorry with turnip lights all around him. Ann and I had been on holiday to Egypt that year and we had brought back a couple of Gallabiyas and head-dresses, which came in useful for the occasion.

The weather was foul, rain came down in bucketfuls, and we got soaked. However, we stuck it out but what a bedraggled looking lot we were.

We sang in all of the pubs in our village, most of which their customers were glad to pay us to get rid of us, because our singing was so terrible!

I thought we did OK, we raised nearly two hundred pounds in four hours, 'but' there again some folk will pay anything to get a bit of peace!

It wasn't anywhere near as much money as we had raised from our other events. However, little fishes were sweet and the money all went into the pot. At this time our funds were in the region of six thousand pounds. Some of the money raised was used for legal fees, which enabled us to fight our cause more strongly, although at times legal advice was not forthcoming.

The Lawyers seemed only to be interested in taking money from the group which resulted in very little or no return for it.

There was a time when T. C. Cornwell applied to Stafford Borough Council for premises in Wharf Road Gnosall, whilst we, as a group, were making protests against such an application. On this occasion we were supported by the Council decision in not allowing the Chemist to go ahead.

During the early part of 1997 after trying to obtain premises in Gnosall the inevitable happened. The chemist did come to our village they finally leased a shop in the High Street about half a mile from the Practice.

Shop fitters came in, and it looked very smart when they had finished it, much to our disappointment. Anyway they were here now. We had to fight tooth and nail to get them out, and that was not going to be easy. The Cornwell Pharmacy opened during February 1997. There was never any confrontation between the Chemist and our group, we were squeaky clean and we intended to keep it that way.

Our fight was to protect our Doctors, and the wonderful service they were and are still providing. The patients could have taken their prescriptions wherever they wished, they had a choice, and their choice was to stay with the Doctors. Ninety eight percent of the patients made that choice. They did not patronize the Cornwell Pharmacy from the day it opened until the day it closed in February 2000, our doctors were able to carry on business as usual, supplying medication and drugs to the patients for the next six months.

The Doctors' health care shop was a complete success. The next move was to try to lease Pharmacy premises, which would enable us to continue to supply medicines and drugs on a permanent basis within a mile radius of the practice. Patients from the surrounding villages and outside the one-mile radius would not be affected. If we could pull this off then the funding of an extra doctor would be able to continue.

It all came together very well, when a Pharmacy became available at Norton Canes near Cannock. It was called Maxim Point, and it opened as a limited company in October 1997. We had already put together details of the possibility of operating a courier service. From there the action group was asked if they would collect signatures from the patients offering a freedom of choice as to where they wanted to shop. Paperwork was delivered from door to door for patients who wished to sign in favour of the proposed courier service, also signatures were taken at the health care shop, and Surgery. Needless to say we were not without criticism, in fact we were heavily criticised by the Staffordshire Community Health Council who claimed that we had used 'bully boy tactics' in our approach towards patients for their signatures, and that we had set up camp outside the Doctors' surgery, the patients almost having to run the gauntlet to gain access. The chairman also claimed in her letter to the Staffordshire Newsletter, dated twenty seventh August, 1997 that she had received complaints from people in Gnosall saying that they felt intimidated by the actions of our group.

When I asked the chairman if she could substantiate the statement she had made to the press, she could produce nothing to show that there had been any mishandling of the situation. It was a complete fabrication of the truth because there was plenty of room on either side of our so called camp for the patients to pass into the surgery premises. The camp was a gazebo and used for the day as a shelter. This caused quite a laugh when one of the patients 'Gordon' sent a letter to the local paper saying that he had sited a seventy two year old bully boy with his seventy year old wife, both were heavily armed with pens and papers. Needless to say that was Ann and me. Support continued to grow, so much so that the action group decided the time was right for us to go ahead and start a courier service, Barrie negotiated the purchase of a white van which was paid for from our fund raising account. From then onwards, it was all systems go, the service was set up and running by the Gnosall Surgery Action Group from the second of October 1997.

The courier service operated from Gnosall Practice delivering medicines and drugs to the patients' homes at an agreed time to suit them. There were as many as two and sometimes three deliveries a day, it was an efficient and a much-appreciated service.

In the February of 1998 a newcomer joined our group, Julie. She was a very intelligent young lady, she worked at our Village Practice, and was a great organiser, she had a brilliant memory and a good personality. We worked together for three years and became very good friends.

Julie looked after the advertising for our "car boot sales", which on a good day we would raise three to four hundred pounds, and we held at least six, sometimes seven boots per year. Our Whist drives were also very popular. They were held purely for community entertainment for all who wished to come along. However the raffle at our whist drive would sometimes raise about eighty to one hundred pounds. Cakes, sandwiches and tea or coffee was provided and included in the ticket price of one Pound.

Julie and Liesa, who worked at the Practice, also raised five hundred and forty five pounds from an Elvis Presley night on the twenty second of July. At this time Ann and I were contemplating holding our second auction which was to be held on the twenty first of March 1998. It was organised by us through the School Headmaster, Mr. Butlin, exactly as we had done during 1996, except that we had much more help available from our members on this occasion. It was another success and we raised almost four thousand pounds in about four hours, a little less than our first auction. Doctor Mick Mulligan was quite concerned about me, just after the auction. He thought that I looked very poorly and suggested that I was doing too much, and that I should take things a little more easy. I made an appointment to see Mick at his Surgery when he asked if I had thought about leaving our cottage and coming to live in the village. I listened attentively because he always made a lot of sense. He put to me the difficulties that could arise if I was to have a serious medical problem. Also the fact that Ann didn't drive would produce considerable difficulty for us. There was a lot to consider here such as maintaining our generators, plus half an acre of garden, and a cottage to look after; our nearest neighbour was half a mile away at the Swinnerton Farm. On top of that we were three miles from the Village. To think that Ann and I had coped with all of this and more, for thirty-four years. Mick could see this in a different light to me. As I left the Surgery his comments were "just a few things for you to think about". I could see the reality in what could arise from the situation.

I welcomed Mick's comments, gave them a lot of thought and then put the question to Ann, who reluctantly agreed that it was the right move for us to make.

We had been very happy there and if we were going to stay we would almost certainly watch the garden go wild, the cottage deteriorate, and although we loved the place so much we could have lived to hate it. The cottage was kept in first class condition. There was always something to do and so for the next three months we worked and looked after our garden, until we decided to put it on the market in May 1998 and it sold to the first person that came to view.

We are friendly with the people who bought it. From time to time we are asked to visit them, they love it as much as we did. They continue to open the outhouse windows for the swallows to return, just as we had done for over thirty years. We bought a house in Gnosall within walking distance of our Doctors and the shops. The house has all mod cons, we now have a video, a freezer, a fridge and it is now fully central heated, will I live to understand it all? Of course I will.

During the autumn of 1997 a proposal was made by our spokesman Barrie Williams that the present title of our group, "Gnosall Surgery Action Group", should be changed. It was suggested to the group members that they put forward their thoughts on the matter.

At a meeting held on the first of October 1997 a vote was taken from the group members, all where in favour of the new title put forward, "The friends and Patients of Gnosall Surgery". It was thought that this was more in keeping with the courier service, and also the future supportive roll that it may play within our community. Although we were known as the action group, this was purely to show that our intention was to take action through the courts in trying to get a change in the Law relating to the supply of medicines and drugs within a one mile radius of a Pharmacy. Freedom of choice for our patients would mean that they could take their prescriptions to any Pharmacy including T C Cornwell in the Gnosall High Street. However their choice was to support the courier service and also the Practice at Wharf Road which had been prescribing and supplying drugs and medicines to my knowledge for a period of about forty years. Prior to this there were Doctors Surgeries in other parts of Gnosall. Over eighty-five years ago I understand that a Doctor Steele prescribed drugs for his patients here. Our patients supported us in every way and could see no reason for change here at Gnosall.

In June 1998 the courier service was sold to Maxim Point Pharmacy limited at Norton Canes, near Cannock, it was seventeen miles to Norton Canes from Gnosall. Prescriptions were taken there and medicines brought back and delivered to the patients' homes every working day. There were two and sometimes three deliveries a day, all of this was arranged to suit patients' needs, and although the courier service was not without its problems it was and is a very successful free service to the patients.

In 1998 our first chairman, Roy Barratt, retired from our group. He was a good chairman and carried the job well for over two years.

Roy's successor in the chair was Professor Owen Ashton. Like Roy, he had been on the group since day one. Owen always gave me the floor when I wanted to say something; I suppose it was my stuttery and stumbley way of making myself heard. I was the oldest member of our group; maybe he was making allowances for me in that way.

Although Owen was responsible for chairing the meetings, he also organised many quiz nights and Spring draws, along with Derek Butlin, who, after retiring from the School in 1999 had come to join the group along with his wife Leslie.

We continued with our fundraising activities, car boot sales, whist drives, quizzes etc. Our bank balance was good and we thought it would be prudent to spend some money. We had no legal expenses to pay at this time. The courier service was going well and our patients had continued to boycott the Pharmacy. It wasn't a happy time for us or TC Cornwell and it was obvious that the position Cornwells were in was not good. There was no way that they could stay in business here for very long.

Unfortunately no one wins in this situation, but we as patients of the Practice had to hold on to what we had got, or loose it. In the mean time, we decided, with our Doctor's guidance that we should negotiate to buy a scanner for the patients.

Helen, our treasurer had a good position at the Hospital, and she was also good at bargaining and so we got a good deal. The Scanner was presented to the surgery by ex chair person Roy Barratt and Doctor Mulligan on Thursday the twenty fifth of November 1999.

Towards the end of 1999 there was a rumour around our village that T.C. Cornwell were moving out of the Gnosall Pharmacy in the High Street.

It turned out to be true. They had decided to call it a day and by the end of February 2000 they had sold to Maxim Point Limited.

The courier service was working from Gnosall Practice to Norton Canes at this time, however after Maxim Point Limited had taken over the Gnosall Pharmacy the courier service worked from wharf road practice collecting prescriptions which were taken to Gnosall Pharmacy for processing instead of going to Norton Canes. Drugs and medicines were collected from the Gnosall Pharmacy by the courier service and afterwards delivered to the patient's homes to suit their needs. On the first of March 2000 the Health Care Shop moved from the Wharf Road Surgery to the Gnosall Pharmacy (Maxim Point Ltd) in the High Street.

Our Group continued to do well and was more involved in fundraising for the Practice. Over the past three years we had bought our Scanner, which was very expensive. We also had to pay for an ultrasonographer to come to the Surgery to operate the machine. We had also bought twelve nebulisers which were loaned out on short-term to patients suffering from breathing difficulties. These were to be returned to the Practice and were sent away to be serviced and replacement filters fitted. Glucometers were also purchased and are in use. The buying of an additional probe for the Scanner was in progress; this was a very expensive item but I am sure that our treasurer Helen will as usual get us a good deal.

The year 2000 has passed. I have almost finished writing my memoirs. I have lived amongst the farming community for most of my life; during

Presentation of the scanner by The Friends and Patients Group to Gnosall Doctors Surgery on 25th November 1999.

those times I have made good friends and seen many changes. In the old days the locals would pass the time of day, they would sit on the sandstone wall and put the world to right, not anymore.

Farmers have changed. Like everyone else they haven't got time to talk; its hard enough for them to make a reasonable living, and then on top of it all in 2001 came the return of Foot and Mouth disease which was the last straw for many.

Ann and I have got used to living in the village, being members of our group has helped a lot. Before we left our cottage we had made new friends in Gnosall and we still have the old ones. We are always working to raise funds for the friends and patients of Gnosall Surgery. It is very rewarding especially when you know that every penny goes back into patient care.

As I look back over the years I see that skills of my trade have almost vanished forever. The making of locks by hand has gone; the old lockies of the past have died taking their skills with them. They never had time to write their memoirs, life was all work booze and bed. Many of them couldn't write anyway, and if they could have, they wouldn't have bothered. I still have my skills taught to me by my father. Thanks to his teaching, this has provided me with a very satisfactory and happy lifestyle.

Appendix

THE RACKBOLT LOCKING SYSTEM

A very expensive system, robust in design, hand made from high quality gun metal castings the weight of which was around ten or eleven pounds each lock.

This type of lock has been made over many years for one of the big five banks. It was fitted to the front door of most bank branches throughout the country. The older type rackbolt lock was operated by turning the knob or handle which resulted in the movement of two sliding bolts, each sliding bolt was drilled and tapped. There were two steel rods which were threaded at one end only, these rods were screwed, one into each of the sliding bolts.

The rods were to secure the doors and travelled in a vertical direction into their respective locations or keeps, one up and one down, after which the key operated the main locking bolt in the usual way.

This type of lock has long since gone except for the odd few, which are scattered around the country. One of the well known lock makers of many years ago was Marley Brothers of Ghost Works. I think it was at Deritend, Birmingham. Another was the Elsey Rackbolt Lock and also the Adrian Stokes Rackbolt Lock. During 1962 I had a visit from Mr. Neville, Worralls Locksmiths, London, with a proposition which was to last me all though the rest of my working life. It was for me to make a new rackbolt lock which had been patented by the Worrall Company. There was to be a refurbishment programme throughout most of the branches of the bank and there were approximately two thousand of them in England and Wales, plus a few more off the Mainland.

The new lock had a three-way follower instead of two ways, which moved the rods and the cross bolt into position with the use of the knob or lever handle. The actual locking system was an insert lock handmade with five or seven lever locks. There were very few keys with two steps together of the same height.

It was ideal for servicing. All I had to do was to remove the lock from the door, dismantle it, take out the insert lock, fit a new one and I could supply as many keys as were required. The old type rackbolt locks had the workings built on the floor of the rackbolt lock body which meant that repair work on site was very time consuming, expensive and highly skilled. For me this work certainly was the icing on the cake. I had got the order to make all the rackbolt locks for the bank for the next thirty years.

At the cottage we had our pre war Petter Twin Water Cooled Generator we nicknamed her "Diesel Dolly". I bought her for twenty-five pounds; she produced all our electricity for the bank refurbishment programme, keeping them safe and secure under my locks and keys, which were made by me in a little old washhouse at the back of our cottage. I had a bench, an old leg vice, a linishing machine, a key saw, and an old bench drilling machine.

I had it all and I looked after it until I retired. Poor "Diesel Dolly" burnt herself out. I overloaded her and like the old van she just couldn't take anymore.